DISCOVERING FIRE SERVICE II
THE OTHER SIDE

JOHN OWENS CEO
Professional Testing Corporation

Archway Publishing books may be ordered through booksellers or by contacting:

Archway Publishing
1663 Liberty Drive
Bloomington, IN 47403
www.archwaypublishing.com
844-669-3957

Because of the dynamic nature of the Internet, any web addresses or links contained in this book may have changed since publication and may no longer be valid. The views expressed in this work are solely those of the author and do not necessarily reflect the views of the publisher, and the publisher hereby disclaims any responsibility for them.

Any people depicted in stock imagery provided by Getty Images are models, and such images are being used for illustrative purposes only.
Certain stock imagery © Getty Images.

ISBN: 978-1-6657-3419-6 (sc)
ISBN: 978-1-6657-3420-2 (hc)
ISBN: 978-1-6657-3421-9 (e)

Library of Congress Control Number: 2022922465

Print information available on the last page.

Archway Publishing rev. date: 02/28/2023

THIRD PARTY AERIAL INSPECTIONS

FIRE—EXPLORING—DEVELOPING --PRODUCTION

Founding of Fire Departments in the New World,
Changing from Volunteer to Paid Departments,
Forming Ladder Companies,
Changing from Manual to Automation, Conflagrations,
Manufacturers of Fire Apparatus, Refurbishing Apparatus,
Aerial Fatigue and Failures, Ground ladder History,
Wood and Metal Ground Ladder Inspection and Testing,
Aerial Maintenance- Research and Schooling, Aerial Trivia,
History of Third Party Aerial Inspections

DISCOVERING FIRE SERVICE II

Picture IRES courtesy of San Francisco Fire Department

FIRE PROBLEMS   **ANCIENT HISTORY**   **WORLD EXPLORATION**

AERIALS/PUMPERS   **FAILURES**   **INSPECTIONS**

GROUND   **LADDERS**   **WOOD METAL**

MAINTENANCE   **TROUBLESHOOTING**   **MANUFACTURERS**

(PRIMARY SUBJECT)
THIRD PARTY INSPECTIONS

Books have been written about many phases of fire service. We have read great stories about difficult rescues and special tributes given to the heroes and first responders, who work so hard in fire service, constantly contributing over 110%. These heroes are both professionals and dedicated volunteer firefighters, who jeopardize and risk without consideration, their lives, every day of the year. These firefighters are on call 24 hours a day to save lives, extinguish fires and rescue their fellow comrades without giving it a second thought. There are situations when they are not aware of the extreme dangerous conditions, but they continue working until the last problem is solved. The stories of dramatic rescues, the lives lost, the lives saved, and the conflagrations that plagued the world all over, are all part of a normal day's work for this special type of women and men. What makes these "special" people perform extraordinary acts of courage; is their "Dedication" to the one job that is in their heart.

There are also many books comprised to illustrate brands of Fire Equipment and Products from the fire industry. Firefighters, collectors, and fire buffs worldwide have written book after book on their personal experience with their favorite type of fire apparatus equipment. Many like to write about a specialty brand product with which they favor or have been associated with.

The Development of Fire Service II, The Other Side, is comprised of multiple characteristics concerning fire, establishing the new world, development of fire prevention, establishing fire departments and equipment. The main focal point of this journal is the Safety of Fire Equipment and Training by Third Party Inspection programs of Aerial Apparatus, which I personally helped developed.

DISCOVERING FIRE SERVICE II
THE OTHER SIDE

Written By:
John Owens
Accredited Dept of Labor/OSHA
CEO PTS, INC.
Engineering
Fire Truck Design Cranes/Derricks

Photo courtesy of ALF

<u>In Memory of:</u>
Bob Rumens
Navy Fireman
Past Chief Mechanic
Virginia Beach FD

<u>For:</u>
Doug Moss
Inspector-PTS, INC.
Chief Mechanic
Virginia Beach FD

<u>For Fire Mechanic:</u>
LES MCCLURE
Chief Mechanic – 50 years
Harrisburg, Pennsylvania

<u>For:</u>
All New York 9-11
Firefighters, EMS
Police Officers

CONTENTS

SPECIAL THANKS TO:

I appreciate all who agreed to be a part of this Journal, and for the diversity of pictures which illustrate the progress of firefighting from past to present.

Fire Engineering Magazine; C. Dewey; Reliance Fire Museum

Fire Departments: The San Francisco, California, Fire Department, The City of Richmond, Virginia, Fire Department and Taunton, Massachusetts, Fire Departments for their pictures

Manufacturers: American LaFrance/LTI/ Emergency-One/ Seagrave/ Mack Trucks/ Smeal Fire Apparatus/Pierce Manufacturing/Altec Neuco/ Greenwood/Maximum/Pierreville

Special Thanks: A special thank you to the Fire Departments shown below who used Professional Testing's inspection program throughout the years. If your department is not listed, PTS thanks you for your support:

Chula Vista, San Diego, CA.; City of Atlanta, Cobb County, Duluth, GA.; Charlottesville, Richmond, Williamsburg, Virginia Beach, VA.; Nags Head, Kill Devil Hills, Duck, Wilmington, N.C.; Jacksonville, Daytona Beach, Dade County, FL.; River Ridge, Jefferson Parish, East Bank, West Bank, Laplace, Kenner, Thibodaux, St. Bernard, Shreveport, LA.; San Antonio, Corpus Christi, Plano, TX.; Wilmington, Smyrna, Dover, DE.; Harrisburg, Bethlehem, Reading, Butler, PA.; Baltimore County, Baltimore, Towson, Pikesville, MD.; Beckley, Huntington, Parkersburg, Wheeling, W. VA.; Gloucester City, East Orange, Tinton Falls, N.J.; Larchmont, Hicksville, Levittown, Ronkonkoma, Bayshore, Sayville, Montauk, N.Y.; Wilton, Bridgeport, New Haven, Middletown, CT.; Woonsocket, Lincoln, Warwick, East Greenwich, R.I.; Hull, Taunton, Middleboro, Bridgewater, Brockton, MA.; Burlington, Essex Junction, Shelburne, Williston, VT.; Lima, Fort Shawnee, OH.; Harvey, Bloomington, Peoria, Belleville, IL.; Colorado Springs, Westminster, Lakewood, CO.; and Lake Oswego, OR. All the cities in Canada

INTRODUCTION:

DISCOVERING FIRE SERVICE II, THE OTHER SIDE, is a Journal exploring FIRE at its earliest known stage to man. The expansion of the world population increasing from the early ages recorded to present day. Establishing new worlds through exploration of different continents. The stages of firefighting and the development of Fire Departments; Volunteer and eventually Paid Firefighters. The stages of Manufacturing Fire Equipment, and a listing of the original and current Manufacturers of Fire Department Equipment. Included is general information about design features, failures, and various situations that overwhelm some types of fire apparatus that were used in firefighting. A summary of the many world Conflagrations. Introduction to Maintenance, (Inhouse and Third-Party Inspection) of Fire Apparatus to include Ground Ladders. Also, Trivia questions about failures of apparatus and equipment.

Fire can be of a friendly, or of a rogue nature. When we have friendly fires, we enjoy the warmth and uses that satisfy our needs. Fire can make us feel safe from the elements. When we are wet, it will dry us. When we are very cold, it will warm and comfort us. Fire will give us light when it is dark. Fire is used for cooking both indoors and outdoors. Campers enjoy sitting outside on cool nights by a campfire. Manufacturing started with blacksmiths, heating coals to form metal products as far back as 1500 BC, friendly or not so friendly. Fire is used in many aspects of manufacturing including steel and power companies. Fire was used in steam propulsion of vehicles. When **FIRE** becomes our **VILLAN**, it is no longer wanted, and a need for control or extinguishment is necessary. When forest fires are out of control, sometimes fire must be used to counteract and slow down oncoming fires until they finally burn themselves out.

Fires have been fought long before BC. Most were probably caused in early time by lightning strikes that had to burn themselves out, as equipment used (if any) would have been antiquated or non-available at that time, and that was how it was.

Volunteers at first fought the fires, then paid Fire Departments were to follow in the later years. As population increased and the west world was explored; colonies, cities and municipalities were under development. There was always the fear of fire, and a means to extinguish was needed. Manual tools, manual pumpers, makeshift- ladders and buckets, were used for many years. Manual operated pumpers were first purchased from England. The manual pumps assisted firefighting until around 1900. Volunteer Fire Departments were formed in the new world, later to be a partner with Paid Fire Departments. In the early 1800's, manufacturers of ground ladders, and manual extension ladders were developing. By the 1870s, multiple section manual extension ladders to ninety feet were being developed. After failures of the extra-long ground ladders, these multi-section extension ladders were reduced to sixty feet maximum. At that same time seventy-five-foot horse drawn tiller, wood aerial ladders were being developed

In the late 1799 and early 1800's steam locomotives were being developed for the coal industry. Around the 1820's passenger type locomotives were also being developed to run on rails to various locations.

Around 1850, one of the first manual and horse-drawn steam engines was developed in the US, and by the late 1800's and early 1900's, an innovation of steam engines emerged from various manufactures using self-driving forces. This was the start of automation, the change from manual operation to full automation, later including the introduction of hydraulic operation of aerial devices. With the vast amount of new style pumpers and aerials, maintenance would become the next critical factor to keep all this new equipment into a good working condition.

EARLY DEVELOPMENT

Early Times

For as long as humans know, fire has been a necessity for warmth, for cooking, for lighting up the darkness, for warding off dangerous predators, and for burning and clearing areas of unwanted waste, and as time progressed, it has been used for steam automation, among many other uses.

Fire made by means other than by natural sources—such as lightning strikes, lava from volcanos, or other natural phenomena—dates way back to the *Homo erectus* period in Africa, about one million years ago.

Fire by percussion (striking of stones or possibly rubbing two sticks together) was thought to date back to the Paleolithic period, but it was not until around 4,000 BC, in the Neolithic period, that signs of fire by percussion was discovered.

Records show that in 10,000 BC the world population was somewhere around two million people. By the time 3,000 BC came around, 7,000 years later, in the Neolithic period, the population of the world had increased to forty-five million. The next recorded population period would be 2,000 years later. This period was known as the Iron age—1,000 BC. By then, the world population would increase to an amazing seventy-two million. Continuing to AD 500 (1,500 years later) the world would see the population escalate to a whopping 209 million. The population now is estimated to be over 7.9 billion people worldwide.

During this period of ancient history, there was little record keeping of firefighting and how fires were fought and extinguished. It was not until around 300 BC that signs of fighting fires with the assistance of a manually operated piston-type water pump was found in ancient Egypt. These signs of fighting fire with the assistance of a manually operated piston water pump are attributed to an inventor named Ctesibius of Alexander. His idea was to find an alternative way to extinguish a fire by the means of a pump that was developed to spray water. This pump was not the most efficient but served as an extra tool to help in extinguishing fires. Centuries again passed, with this being one of the first and only means to assist in firefighting. Then around 100 BC, a man named Hero of Alexander changed the design of Ctesibius's manual water pump to improve the performance; he created a manual piston-type water suction pump with flexible hoses to assist. This manual pumper had limited range because of the length of suction and the flexibility of the hoses. It was not until 1540, when a small handheld fire extinguisher was introduced to extinguish small fires. This piece of equipment went on to be the best discovered, and at that time, it was the only thing on record developed to assist bucket brigades along with volunteers and whatever tools used to assist putting out fire.

The population was escalating, wars and skirmishes were everywhere, but firefighting seemed to remain unchanged for the next century, with no introduction of any new type firefighting equipment.

Going back to Rome, around 24 BC, Roman Emperor Augustus, was credited as the father of organized firefighting. He formed a group of local volunteers called "watchmen," to keep a visual watch throughout the city. The watchmen would sound an alarm the minute a fire was noticed. This seemed to be the routine followed anywhere firefighting was needed, until AD 60 when in Rome a first ever fire brigade was formed by Crassus of Rome. This fire brigade had the use of around five hundred volunteers. They were the only ones who could be used to extinguish fires. Crassus had a plan for his future wealth. When there were signs of a structural fire or a building was engulfed, Crassus would be notified of the burning situation before his fire brigade would attempt to extinguish the fire, then he would contact the owner of the property and make an offer to purchase the engulfed building from the owner at a very ridiculous price.

If the owner refused the offer, Crassus would have his brigade leave the scene and let the structure burn to the ground, without any attempts to save or extinguish.

After a few years later, Roman Emperor Nero caught on to what Crassus was doing, and to eliminate Crassus's monopoly on firefighting, Nero came up with a different idea—he put together a group of volunteers called Vigiles. These Vigiles would not only fight fires but they would also act as Nero's police force.

In AD 64 Rome suffered its most serious fire that destroyed nearly two-thirds of the city. Theory was that Nero had planned this massive fire, but he was out of the area when the fire destroyed the city, so who could blame him for the arson that remained unsolved?

Wars and skirmishes continued throughout Europe and the eastern part of the world for many centuries to come.

Ways to create fire were always thought of. How the many arson fires before BC were ignited is a mystery. It was around AD 500–600 when Chinese started inventing "fire inch sticks." In the next 700 years the Chinese improved on this method substantially. Years later, around the 1820s, the first stick match with sulfur and gum was introduced. This match was short-lived because it was too dangerous to light as the gum flame would fall like candle wax. In the 1830s, white phosphorus stick matches were invented—the kind you could strike on your jeans to light or with the use of your thumbnail. Then an innovation of matches came around 1890, when paper matches in folding cover books were introduced into the market. From those days to the present day, matchsticks and paper books have changed to propellant handheld lighters. Though matchsticks are still available.

In the year 1488, ships from different countries started exploring and discovering new and different ways to find riches and land. The Portuguese ships found a new direct route to India, traveling around the south coast of Africa, while the Spanish fleet in 1492 traveled to the Atlantic West to what was thought to be the end of the world—where ships would fall off the edge of the ocean. There were three Spanish ships under the control of Christopher Columbus, which would find land in the Bahamas, and other Spanish ships that would travel south around South America, trying to find a shorter route straight through the Americas from the Atlantic Ocean to the Pacific Ocean and up the western coast. There was no short route from the Atlantic Ocean to the Pacific Ocean through the Americas until the Panama Canal was completed in 1914.

New York

Europeans had little care for the Americas and from 1520 to 1648 were in a series of religious wars. The only ships from Europe to sail to the west were ships controlled by freebooters (pirates or lawless adventurers), who would soon discover different areas of North America.

In 1524 an Italian explorer Giovanni da Verrazano would be the first to discover New York Harbor area. There was no settling until later when the Englishman Henry Hudson would sail up what is now called the Hudson River. Years would pass before new explorers would arrive.

It was in 1624 when Dutch explorers would land and build a permanent trading post on the southern tip of Manhattan Island, which at that time would be called New Amsterdam. Then twenty-four years later, in 1648, the Dutch started a settlement known as New York City.

Roanoke Island

One of the first of two European expeditions by Sir Walter Raleigh sailing toward the West was in 1585. Sir Walter Raleigh landed on Roanoke Island (Dare County, North Carolina). Before continuing on his exploration, Sir Walter Raleigh appointed as governor of this Island Mr. Ralph Lane, who was to establish a colony of 180 volunteer settlers on the island. Two years had passed by before a second voyage, early in 1587, was to return to Roanoke Island with supplies for the newly established colony. After landing on the island and searching it, the original 180 settlers from the first voyage were missing and no signs of their whereabouts was ever found.

The newly appointed governor, Mr. John White, from the second voyage, was to establish another new colony on Roanoke Island with settlers from the second voyage. Supplies were low. Later in 1587, Governor White would return to England for additional supplies for the new colony. Governor White would leave behind his wife and granddaughter, Virginia Dare, whose name was used for Dare County, NC, and for Virginia on the eastern coastline of America—which later became the state of Virginia. This colony would have to live on the supplies they had until John White returned with new supplies.

Governor White did not return with the new supplies for the colony until three years later, in 1590, only to find that, like the first voyage, there were no traces of his family or any habitants of the second colony. Thus, the "lost colonies' of Roanoke Island remain a mystery.

As all persons become missing from both voyages, and this island is in a hurricane area. Could this be a reason no evidence could be found?

Jamestown

In 1606 King James granted a charter to the English Virginia Company to establish a settlement in North America. The English fleet composed of one hundred all-male settlers and three ships, *the Godspeed, Susan Constant*, and *the Discovery*, would eventually locate in the Chesapeake Bay. Commander of the sea voyage, Christopher Newport and Captain John Smith formed a counsel to discuss a suitable landing area. On May 14, 1607, they landed on a narrow peninsula, an island in the James River where they departed from the ships. The settlers built a triangular shaped fort, it was named James Forte and other names. In the summer of 1607, Commander Newport, went back to England to load up on supplies for the colony, as there was widespread starvation, along with the many skirmishes with Indians. Two years passed and Commander Newport had not return with any supplies for the colony. In the summer of 1609, Captain John Smith, decided to return to England for supplies. Captain Smith left the colony for England and the inhabitants of the colony had to forge out the next harsh winter with very little food. This period was called "The Starving Time" where nearly 100 perished. In 1610 the remaining colonists were about to leave the area when two ships returned with supplies and 150 new settlers (148 male and 2 women). John Smith stayed in England and did not return.

Sir Thomas Dale was to take charge of the settlement. He enforced new laws to restrict interactions between the settlers and the Algonquin Indians. He made skirmishes with the Indians, raiding their village, and burning down their shelters. In 1611 new buildings were going up east of the fort, which later would be called Jamestown the Capitol of Virginia. There would be many fires in Jamestown, there is no evidence of the use of any type firefighting equipment in the early settlement days, other that the bucket brigade and whatever modified tools to help rescue and extinguish fires. This was still a full volunteer effort. Later in 1699 the Capitol of Virginia would move from Jamestown further inland to Williamsburg.

PLYMOUTH:

Ships continued to explore the west Americas, English, Spanish and Portuguese to name a few. Many carried supplies back and forth to the new world. Some ships carried slaves, but most carried supplies and pilgrims to explore the new world. In the year 1620 a ship from England, the Mayflower, and a ship from the Netherlands, called the Speedwell, were hired to take food and supplies, along with Pilgrims, to the shores of Northern Virginia.

The Speedwell after meeting up with the Mayflower in Portugal was taking on water and was not ship worthy for the long voyage. Repairs were done, but in the end the Speedwell was not able to sail the voyage. Supplies and the Pilgrims from the Speedwell was then transferred to the Mayflower and the voyage to Northern Virginia would begin. The voyage was set to arrive somewhere between Northern Virginia area all the way to the Hudson River. That was the original destination for the voyage. It was some 66 days of riding violent storms and rough sea swells heading west to the New World where land was eventually sighted. It was Cape Cod. They knew this was not the intended destination area they planned to port, so turning southward they continued to head for the Hudson River. The seas got very rough heading south and the Mayflower nearly capsized, so a decision to turn back and landed at Provincetown Harbor on Cape Cod was then decided. Exploring the area in late December 1620, the Pilgrims also decided that Plymouth would be the destination to start building a new life in this New World.

This would be known as the town of Plymouth. Plymouth holds the great distinction of being one of the first and oldest settlements, along with Jamestown in Virginia.

Ships from England started to arrive in the Plymouth area. In 1630 Puritans arrived called the Massachusetts Bay Colony from England. The Puritans purchased land from the Plymouth Colony, to form the town of Boston.

BOSTON:

In 1631 the town of Boston was increasing in *size a*nd had developed the first Volunteer Fire Department in the new world. Years later and expanding in 1653, the town of Boston purchased their first-hand operated fire pumper from England. Boston firefighting would remain the same procedure until 1678, when Boston would develop their first paid Fire Department. Then years later in 1711, Boston would develop a group of fire wards, responsible for the operation of firefighting and maintenance of equipment all around the Boston area. In 1799 some 88 years later, Boston would purchase from England the first leather hose to assist their manual pumpers.

There was a new telegraph alert system invented to install in all of Boston's fire wards. This new Telegraph Alert System was developed in 1851. By Early 1852, this system was working and was transmitting alerts in all of Boston's fire wards.

Boston had several manual drawn ground ladders and hose carriages, with hand push-pull operated pumpers. It was not until 1910 when Boston had their first motorized apparatus and Steam Engines. As years passed by from 1914 to 1923, new horse drawn equipment and motorized equipment was utilized and the manual drawn ground ladder and hose carriages were retired. In 1926 the last of the Steam Engines were retired and Boston was a full Motorized Fire Department.

ESTABLISHING LOCATIONS:

Ships were still landing, immigrants were migrating from various countries and coming to the new world to form towns, cities, villages, colonies, parishes, boroughs, and municipalities. This was not an easy venture for the migrators when first arriving at the Americas, facing the harsh elements, starvation, sickness, and adopting to the local environments.

Migrators were on the move establishing locations, like Philadelphia in 1736, on the west coast, ships landed in San Diego in 1769. Hoarse Greeley stated go west young man, New Orleans developed in 1829, Houston Texas 1837, both had opposition. Sacramento and

GOLD discovery in 1839, Los Angeles in 1871, with a gold rush. Gold was also discovered around Denver in 1858.

CONSTRUCTION:

After long journeys, facing whatever problems that would arise, the migrators would settle down to their destinations and construction of shelters would begin. Various denominations and types of settlers found areas that suited their individual religions and beliefs. As more buildings appeared, communities started to come together. Business offices were constructed, which became the downtown of the community. Now there becomes a permanent threat to each settler and business and that is the possibility of FIRE.

As the community forms, there must be some type of government control with representation from the locals to administer law and order and make decisions for the community. The townspeople would have to elect a Governors, Mayors and appoint Commissioners, Fire and Police Chiefs and whoever else it would take to help run their community.

VOLUNTEER FIREFIGHTERS:

Volunteers Firefighters have been known to fight and extinguish fires from as far back as 24 BC, with Roman Emperor Augustus and his Watchmen. In this New World volunteer fire fighting started around 1600. There was minimal equipment the volunteers could use at the start. Buckets, makeshift ladders, and whatever means volunteers could find to knock down a fire. The only manual pumpers and handheld fire extinguishers had to come from Europe and if the community did not have funds and was not large enough, this was not an option. Except for the larger communities that already established, like Boston, in 1678. Most paid Fire Departments took longer time to transfer from volunteer to paid and paid departments still utilize volunteer status. Volunteer firefighter's both men and women, are still the backbone of firefighting. Present day, there are more volunteer firefighting communities than paid cities, in the U.S.

As settlements were developing, one of the natural disasters was fires, caused by natural sources or human error, accidental or on purpose. At this period, when a bell was ringing, or volunteers ran through the streets, yelling "FIRE" to alert the public was the only way to get the firefighters to the site. Fires were extinguished by Bucket Brigades at first. Lines of Volunteers would start filling buckets of water at sources already established and pass the buckets from person to person, to pour onto the fire, until the fire was extinguished or contained. Wooden Ground Ladders and roof hook ladders were used in many cases for rescue and to give better access to extinguish roof fires. These ladders were stored in areas for ease of access when needed some were hung on public fences throughout the settlements so the volunteers could grab them quickly on the way to the fire. Townspeople were always working together, trying to keep what they slaved so hard to create, from burning to the ground. Volunteers were doing everything they could to rescue victims and supply water to extinguish the fires with limited equipment available. One example like the fire in Rome, the city of Miami, in 1896, half of the city was deliberately set on fire, by the townspeople just to protest high insurance rates taxed on them.

In the mid-1700's, one of our founding fathers, Benjamin Franklin, our Sixth President, was credited as being the first to start a volunteer fire department in Pennsylvania. Today, volunteers still play a major role in firefighting and assisting their departments with their overpowering dedication. Volunteer firefighters are used in most U.S. States and throughout the world. There are over twenty-nine thousand fire departments in the United States, most of them volunteer. Worldwide, fire departments have both paid firefighters and volunteers. In the U.S., fire departments are still using a combination of volunteers along with paid firefighters. In Louisiana, there is a Parish with over three hundred volunteer firefighters; and many other states, from Long Island to the West Coast, which have volunteers for firefighting and rescue.

A central headquarters was established to send teams of volunteers to various locations. It was difficult to find a certain team to dispatch and at times they would have to pay or bribe a volunteer to make up a team.

Firehouses had to be modified when horse drawn ladders and pumpers, or rescue equipment started to show up. Firehouses had wooden floors which had to be modified for the horses, as a safety measure, also rigging and harness brackets made for quick hookup. One of the first ladder truck was pulled by an oxen and did not respond to fires quickly.

Picture courtesy of Fire Engineering

Many of the older Firehouses had narrow door widths, basements, wooden floors and had to be modified for the future apparatus. All Volunteer fire companies and departments, have a chain of command administered so everyone will be on the same page when a fire breaks out and their responsibilities known.

THE MAKING OF A FIRE DEPARTMENT:

Communities all over the country were developing into a large Metropolis. As the cities were enlarging more fire protection became evident. It would take more paid or volunteer firefighters to service the area. A chain of command would have to be administered to handle the expansion. The Mayor or Commissioners would have to select a Chief to oversee the paid Fire Department operation. Volunteer departments would vote among themselves, or the city managers could decide the chain of command. Different positions would need to be filled in the command chain. Along with firefighting, purchasing of equipment to assist the firefighters would be one of the most important positions in the department. Until the 1830's, firefighting equipment sources throughout the world was very slim and hard to find. Training of personnel for various positions, such as Chief, Lieutenant, Manager or Firefighter, was usually based on experience OJT, (on the job training), they might have to fill the different positions.

MANUFACTURERS:

In 1832 one of the first made in American fire apparatus manufacturing companies called American La France appeared. They started in the manufacturing of various types of equipment. Manual and horse drawn fire apparatus, manual pumpers, and other firefighting equipment was their main products.

American La France continued this similar type of manufacturing until the year 1873, when American La France was to form another company called Truckson La France. This division would begin building horse drawn apparatus, containing wood manual ground and extension ladders. In 1903 Truckson La France would form American La France Engine Company in Toronto, Canada.

The late 1800's, was the beginning in the New World of manufacturing fire equipment. Different ideas were appearing on the drawing boards everywhere. In 1886, a San Francisco, Fire Department Superintendent Daniel Hayes was designing a fire apparatus aerial wood extension ladder that would reach up to 70 feet in the air.

Prior to his design, manual ground ladders would be tied together to reach higher footage. At that time, although it was not a good idea, that was the only way to reach a higher elevation to fight fires or to do a rescue. This new wood 75-foot aerial ladder was mounted on a horse drawn chassis with a tiller driver. This ladder would have manual extension and manual rotation and manual screw elevation. Daniel Hayes and American La France would merge and work together building over two hundred ninety wooden aerial ladder apparatus. These wood extension ladders would have their problems with failure. Maintenance to preserve the wood quality and strength was a requirement. Then in the early 1900's light duty steel type aerial extensions ladders were introduced. It wasn't until the middle 1950's, when most of the wood extension ladders were removed from service.

Another Manufacturer starting to provide a full line of fire apparatus in 1881 is called Seagrave Manufacturing. They also had a manufacturing service in Canada, called King Seagrave. In the 1900's, Seagraves, would discontinue manufacturing wood aerial ladders, and introduced a new style lattice-type light duty steel aerial ladder.

Wood aerial ladders in the late 1700's and 1800's served a primary function, but when over stressed, many failures would occur. Wood extension ladders were very heavy. The wood ladders would have to be maintained constantly, to preserve the wood. Wood aerial ladders were slowly being phased out with the new light duty steel aerial extension ladders replacing most. The wood aerial ladders with tillers, were still in service as late as the 1950's and wood ground ladders in service into the 1980's. Many fire departments wanted to stay with wood ground ladders because of the electrical resistance that steel ground ladders did not have. Early 1900's; as the new "American Made" steel type lattice extension and ground ladders were introduced, other manufactures of similar equipment would emerge with their different style firefighting apparatus and equipment. Many of the new style steel aerial extension ladders failed when operated incorrectly. In the fire apparatus industry, there was a journal called, "Fire Service Ladders 1952," composed by Mr. Roi B. Woolley, a retired-paid volunteer fireman, and Assistant Editorial Director of Fire Engineering. There were no third-party independent testing companies at this time, so the fire department personnel performed most of the tests. This journal showed testing of early manufactured truss-wood and metal aerial ladders; also, wood and metal ground ladders. The journal showed the early versions of wood and steel aerial ladders in full operation and stressed until failure became the result. Many of the tests were with water flowing 90 degrees away from the extended ladder sections, with the nozzle strapped to the tip of the fly ladder section. This test also showed how far the ladder sections would deflect from nozzle reaction just before failure. These tests were only performed to show that the aerial ladder sections could not resist the nozzle reaction when flowing water, thus failure was the result.

On September 23, 1952, a Sub-Committee for the Committee on Fire Apparatus of the National Fire Protection Association met in Cleveland, Ohio, to review the problems associated with testing aerial ladders. This journal was used by the NFPA to publish the latest version and addenda of the current testing procedures of aerial ladders and ground ladders for that time, and this version was recommended at that meeting. Such procedures have been reviewed and renewed, year to year, from fire department and manufacturers' recommendations, and written in new NFPA pamphlets.

AS Fire departments, both Paid and Volunteers, are developing, facilities for training new recruits and firefighting professionals will need to be established. Management and a chain of command must also be selected for the operation of the department and training. As a vast assortment of fire equipment and apparatus became available and more sophisticated, an experienced purchasing department or agent would be necessary. Next, we will need to have a service and maintenance department. This is a critical division to firefighting. It was found that only general maintenance and service to the equipment was performed most of the time up and to the 1980's. Critical items were often overlooked and there were many failures and breakdown to the equipment. As this seemed to be the proper way to have somewhat of a chain of command, there was then, and still is, the bickering back and forth between the senior and the junior authority, when maintenance decisions are made.

Townspeople, Mayors, and Commissioners:

Immigrants migrated from all corners of the world to establish villages, colonies, towns, boroughs, and parishes in this new undeveloped territory called America. Along with all the skirmishes and fighting from the different denominations to establish and take control over areas throughout the New World, firefighting was still a critical issue to consider. The townspeople would have to elect or appoint mayors and commissioners. Volunteer firefighting was, for centuries, the only way to rescue lives or extinguish fires. As cities grew larger and larger, establishing a professional fire department supplied with firefighting equipment, and hiring personnel, such as, fire chiefs, captains, lieutenants, as well as any firefighters it would take to manage the department, had to be implemented. The jobs got much tougher, not only in fighting fires and making rescues, but also finding the right personnel to handle purchasing the equipment and supplies for maintaining and servicing the equipment. Forming the department, recruiting, and training firefighters, and keeping a budget necessary for firefighting and rescue, was a list that had to be carefully planned. As years passed, it became more obvious with responsibilities, liabilities, and compliance requirements.

Purchasing Agents:

Their duties were to locate, inquire, and purchase the best equipment available at the time, for the protection of the firefighter, while trying to stay under the cities' budget. They continually had to research and talk with manufacturers about their unchanging circumstances and assist their own departments to purchase the most universal equipment needed at that time, for them to perform rescues and keep their firefighters safe. One problem the agents encountered was when a manufacturer's representative would convince a department to purchase a piece of equipment that was overqualified for their department's needs, which caused unnecessary expenditures, just to get a piece of "vital" equipment into the firehouse. Also, the fluctuating economy is one of the biggest headaches for most purchasing agents. Each year the budgets would change, and it became more difficult to balance, as the cash flow disappeared. Commissioners, Governors, Chiefs, and Budget Planners are faced with this problem every year; and unfortunately, fire and police departments always seem to be on the top of the budget 'chopping block' list. After making the budget, trying to stick with it when emergencies happen is very difficult. These annual budget cuts will prevent the purchasing agents from upgrading or even purchasing new equipment for the coming year. There are times when parts to repair equipment cannot be purchased because the city has run out of funds. Some cities would have to close fire stations and layoff firefighters just to balance their budgets.

There are departments that had to sideline equipment because of the repair costs. It's a never-ending battle for these professionals to keep firefighting in prospective.

Maintenance:

Mechanics play a very large part in firefighting. They must be capable of handling the daily service of the equipment, performing safety inspections on a regular basis, doing emergency repairs on the equipment when called upon, keeping the equipment running, and furnishing supplies to locations when involved in emergency situations. They accept the delivery of new equipment, and some mechanics must go out of town to do final acceptance at the manufacturer. Mechanics must be knowledgeable and trained on various types of apparatus from the front to the rear. They are on call twenty-four hours a day; and most of the time, the situations the mechanics must be ready for are not always easy tasks. Maintenance is hard to keep intact. Budget cuts create layoffs, and shortage of supplies. Senior mechanics must fight tooth and nail with their hired mechanics, as do the Department Chiefs and Captains, just to keep enough skilled labor to run a daily routine workload. Buyouts and retirements sometimes eliminate the passing of years of experience to their understudies. Layoffs and downsizing make keeping up with repairs and proper inspections very difficult. These mechanics always are under a stressful situation and demand. They deserve a 'hand of appreciation.'

Equipment Manufacturers and Suppliers:

The demand for specialized equipment brought competition to manufacturer's specially designing equipment to meet the needs for firefighting and rescue for the fire departments. In the early 1800's, many venders started manufacturing in the New World. Before that time, most of the firefighting equipment had to be purchased from Europe or was made makeshift to help meet the needs. As towns, cities, boroughs, and villages rapidly grew, there was always a new request for the manufacturer to come up with a better means to solve the problems in firefighting and rescue. From the early years of wood to the present-day metal, equipment went through the trial-and-error operation process. Many times, when failure occurred, there would be no solution to the problem. Injury and failure resulted. Many experiments and tests of the products were performed. Extension ladders were tied together to reach higher locations, and that was never a good idea. In 1886, a San Francisco Fire Department Superintendent Daniel Hayes came up with an idea to manufacture a fire apparatus mounted wooden aerial ladder. His first manufactured aerial was a seventy-five foot, two section wooden ladder with manual extension, manual elevation, and rotation, and was on a horse-drawn chassis with a tiller operator. This ladder was first purchased around the mid 1890's. By the mid 1920's, there had been so many failures of the wood extension ladders, that fire departments sent out a request for manufacturers to find stronger material for aerial ladders. With persistent research and development, products became increasingly advanced. Manufacturers were constantly upgrading for safety and efficiency. Steel Stainless and aluminum were the next materials to be used in manufacturing aerial ladders and ground ladders.

We are now in an age of computers. They not only assist in engineering, but have universal knowledge, and are used in manufacturing, and the operation of equipment. It's an enormous evolution from the early years of manufacturing. We thank the men and women in the manufacturing fields for their contribution to the advancement in firefighting.

In-Fire House Safety Inspectors:

Unlike general maintenance, keeping the equipment in good working order always requires some type of in-house inspection on a regular basis, to make sure the equipment is maintained, and serviceable. These periodic inspections are usually performed by the mechanics, or a special in-house inspector. When there were inadequate skilled department personnel to perform in-house inspections, the manufacturers were called upon to assist. Sometimes, the equipment had to be sent back to the manufacturer for service and inspection. This was a very costly venue. Training and educating personnel to operate and service the equipment was very crucial. When fire departments had the personnel to perform these duties, operation, maintenance, and inspections would seem to flow without many problems. Sometimes, fire departments with low budget or minimum operating capital would have to move the qualified inspector/mechanic to a different location, or the individual retired, and their training expertise did not get passed on to their replacement. When this happened, the inexperienced inspector/mechanic was working on thin threads, and the equipment had a much greater risk for future failure or breakdown. This did happen on many occasions, where numerous problems with the equipment went unnoticed. The personnel using the equipment would have to assume there were no problems when the equipment was returned to service.

THIRD PARTY INSPECTIONS:

In 1975, a new business was introduced to fire service, called Third Party Safety Inspections of fire apparatus, ground ladders and pumpers. Third Party Inspection would be a good assistant to maintenance and training, also to assist the manufacturer of fire apparatus and other products. An additional source of experienced inspection of the aerial equipment has helped limit liability and failures of apparatus, and to assist fire departments service technicians on periodic inspections and maintenance.

Third Party Inspections of Fire Apparatus was never a requirement, only general service until the 1950's, when the NFPA developed service recommendation pamphlets for fire department in-house testing of aerials, ground ladders and pumpers along with other products. These fire apparatus inspections were usually performed by the fire department personnel or periodic manufacturer service. The procedures were recommended for manufacturers, or the fire department to perform on a regular basis, but not as a mandatory requirement. There have been many accidents deaths and failures connected to firefighting equipment, even as late as the 1970s, Fire department apparatus was mentioned in General Industry Standards Safety and Health Standard 29 CFR 1910.67-(b)-3, Vehicle-mounted elevating, and rotating work platforms, but was exempt under these rules. This OSHA standard exempted fire apparatus from inspection and testing, except that the vehicle the aerial is mounted on must comply to standards. The labor department did not want to interfere with local municipal government fire equipment investigations.

In the early 1970's, I was working for a non-destructive x-ray and special testing company with multiple business features. My job criteria, was a certified crane and derrick inspector, along with being certified level 2, in NDT inspections, including MT, UT and PT. This company was certified by Cal. OSHA for crane and derrick inspections, and by the Atomic Energy Commission for having the largest x-ray source in the Western Hemisphere. They offered x-ray, magnetic particle, and ultrasonic inspections that were performed at power plants and for manufacturers with large ferrous castings, also for corporations needing special weld inspections. Crane and Derrick inspections were performed at Municipalities and for Utility Power Companies on location.

There were a few local fire departments that requested our services for a second opinion to validate apparatus problems, before sending the unit back to the manufacturer. Most problems were like situations we had found with cranes and derricks previously inspected, so we were able to isolate and document many of those problems, and then have the fire department notify the manufacturers.

When working with fire departments, there were many opportunities to have the adrenaline rush while riding with the fire chief, the lieutenants, or the skilled mechanic to a working fire or emergency. You never knew what to expect until you got there, always gearing up for the worst situation, but hoping it to be just a minor incident. We had a chance to see how the equipment was set up and functioned at an emergency and where the limitations of the equipment would restrict the firemen from performing their tasks. Contacting fire departments for many decades, explaining, and recommending how they should research equipment best for their situation, assisting departments with writing future specifications for equipment, working with manufacturers as an independent Third-Party inspection company, inspecting apparatus in both the United States and Canada, and investigating accidents and structural failures were some of my experiences.

In 1975, at the Denver, Colorado Fire EXPO Convention, as an accredited crane and fire apparatus inspector working for and representing a non-destructive X-ray and inspection company based in Southern California. I introduced to the fire industry a new independent inspection program for fire apparatus safety that utilized and exceeded the NFPA Pamphlets and their recommendations. This independent program included a thorough visual inspection of fire apparatus, a non-destructive magnaflux inspection of welds, ultra-sonic inspection of bolts and pins and load dynamometer tests for strength and stability. A Special Test, the use of dial indicators for extension and elevation cylinder drifting tests was not included in the original package and added later. After PTS was incorporated in 1978, I developed a 'Special' testing procedure called "Cylinder Drift Down" with dial indicators. This procedure determines how much the tip of a fully extended aerial ladder or platform could lower, within thousandths of an inch, when elevated to any degree of elevation and to any length of extension, just from the internal leakage of the elevation cylinder(s). This was a new procedure that proved to be very successful and is one-hundred percent accurate. Manufacturers accepted this type of testing and issued recommended cylinder drift, for their product cylinders. The results are conclusive that a dial indicator cylinder drifting test did show that hydraulic cylinders do leak internally, so infinitesimal, and could make an enormous lowering of the tip of an extended aerial ladder or platform, with just a few thousands of an inch internal leakage.

This Convention was successful, and from that date, Third Party Inspections were introduced to the Fire Apparatus and Ground Ladder Industry. At the beginning, there was pro and con controversy concerned about removing apparatus from service. Chiefs were concerned more that they would not have a piece of equipment to send to an emergency rather than if the equipment was safe, or if there were any unknown problems. The Chiefs were also concerned how long repairs would take if the equipment was removed from service. Many of the larger fire departments were concerned about the number of units that could possibly be removed from service and were reluctant to start with this program. This was a period where medium duty aerials were abundant, and the newer heavy-duty aerials were still on the drawing boards.

Later in 1975 this inspection program was sold by the company I was originally with, to a fire apparatus salesperson who developed a new inspection company using these inspection criteria. A short time later the salesperson offered the inspection program to Underwriters Laboratory. U.L. already had a pumper inspection program and they added this program to their business without having any experience in aerial inspections.

We now have a new inspection service without qualified personal inspecting the apparatus. There are no agencies certifying fire apparatus inspectors. U.L was doing pumper flow certifications yearly as per NFPA recommendations. This new aerial apparatus inspection program got confused by many fire departments and manufacturers as a U.L. designed inspection program.

Up until the 1970's, there had been many failures of Fire Apparatus, both aerials and pumpers, along with ground ladders. Most of these failures or accidents were blamed on operator error, and or manufacturer defect to the apparatus or some other equipment cause. Many of the failures were left unsolved. Possibly if a Third-Party test was involved, many of these aerial failures or problems could have been prevented.

It takes a special person to have enough knowledge of fire apparatus and NDT inspecting. From 1972 till 1978, I was accredited by CAL-OSHA for cranes, derrick inspections, also to include fire apparatus in the southwestern states, and introducing NDT level 2 inspections to these inspections. My experiences also consisted of the design of the 75-foot Mack AerialScope and later the 50-foot airport Snozzle with an additional engineer. I have designed many cranes, derricks, and cherry pickers for tree service and power companies since I was 21 years old. I assisted in the development of this Fire Apparatus Third Party Inspection program.

In 1978 I moved from the west coast to the east coast to develop a Fire Apparatus Inspection Company, in Virginia Beach, Virginia, called (PTS) Professional Testing Systems Inc. There was a larger quantity of fire apparatus on the east coast, because of population. I located PTS central between Florida and Maine. I was accredited in 1978, with a United States Department of Labor OSHA General Industry Standards 29 CFR 1910 certificate of accreditation for crane and derrick inspections throughout the U.S. and the accreditation was issued for the next 20 years. OSHA General Industry Standards 29 CFR 1910 accredits inspectors to inspect cranes or derricks from different sizes and capacities. Fire apparatus aerial devices are designed very similar to truck mounted cranes and aerial manlifts, and with this Department of Labor OSHA General Industry Standards 29-CFR-1910 accreditation, it should suffice to show the creditor has the level of experiences to also be an inspector of Aerial Fire Apparatus. Additional Level -1-2 certifications of NDT equipment used in the inspections were also a part of the inspection. There are schools and training facilities all over that accredit inspectors in this profession.

In the early 1980's, there was the big change in aerial ladder design, from light-duty, medium-duty aerial ladder to a more heavy-duty aerial ladder design. Soon all fire apparatus manufacturers would be upgrading their designs to fit this new standard practice. This is when Third Party inspectors and final acceptance inspections began working at the manufacturer's facilities. Third Party Inspection companies started to emerge, but only a few. They included firemen from various departments trying to start into this type of business because they have worked around fire equipment. The knowledge and experience at the early start of these new outside companies was not what it should be. Most used a similar inspection report like I had developed, or just followed the NFPA pamphlet but most inspectors did not have the knowledge or experience or proper accreditation. Manufacturers were upset with the way inspectors are preforming inspections in a short inspection time, and how the many items of the inspectors' notes that did not relate to failure or needed repair to the apparatus.

As time passed, Independent "Third Party" Inspection Companies began assisting fire departments in writing specifications and accepting delivery of new equipment at the fire departments and at the manufacturers. The additional inspection techniques on the apparatus helped find hidden defects, which were sometimes overlooked at normal in-house inspections. This new type "Third Party" Inspection Program was to add additional procedures to the normal in-house inspection, resulting in a more 'in-depth' inspection. This was clearly a new type of fire service business. As I previously stated, in the beginning, it was extremely hard to convince Fire Chiefs how this type of inspection would help lower apparatus failure and repair expenses, but the end results showed conclusive proof that this type of service does works

Not all manufacturers used this type of inspection before final delivery of their equipment. They would simply have fire department personnel come to the factory and go through an acceptance program. Many questions would arise as to the experience of the "Third - Party" Inspector. As one manufacturer CEO stated, and I quote, "Who Tests the Tester?" There are no apparatus training schools. There are NDT schools, fire science colleges, and manufacturers with in-house training that can familiarize inspectors regarding the vast amount of equipment that is owned world-wide. It takes many years of 'on-the-job' training to learn and gain the necessary experience and confidence to perform a systematic "Third Party" Inspection on all types of fire apparatus, and to gain knowledge about the design of the equipment, and to know of their many quirks. Most of the aerials are similar in design; and over a period, one will become familiar with the basic functions of the equipment, and where to look for each problem area on each apparatus.

Years passed before the "Third Party" Inspection procedure was adopted by the NFPA and recommended annually. This was the beginning of many more "Third Party" Inspection Agencies. These inspection agencies were comprised of retired firemen and other testing companies, who hired NDT students, who absolutely had no training in the fire apparatus field. This also brought competitive bidding for apparatus inspections at fire departments. It wasn't how meticulous the inspection had to be, but how many units the inspector could do in a day. Purchasing Agents do not inspect the apparatus, so they would try to get the most out of their 'budget' monies.

Field tests proved to be more complex for accuracy due to paint, dirt, and grime on the equipment. The most difficult NDT inspection of aerial apparatus was in the wintertime when temperatures are around forty degrees or lower. During that time, field testing is extremely tricky. An inspector can get more false readings in the outside cold, than if he was working directly with the manufacturer inside the plant.

Don't ever be under the assumption that everything is OK once an inspector signs off on their inspection report. I have had the opportunity to recheck apparatus previously inspected by other inspection companies. Many inspections showed there were critical areas that remained unchecked and failure of one type or other was evident. Chiefs and Mechanics were very upset when I showed them the problems missed from the previous inspections that still existed. The apparatus was front line and had to be removed from service for critical repairs. The non-destructive testing (NDT) portion of the aerial inspection procedure gives more ways to determine problems in the field, if done properly.

Special Note: It is not the primary duty of a "Third Party" Company Inspector to determine any manufacturer's procedures, nor to make any final decision for the repair of any type of equipment. He/she only is to denote problems if any, report all findings, and let the fire department work with the manufacturer to solve those problems. "Third Party" Inspection Companies should only issue a 'Certificate of Compliance,' which states at the time of the inspection and testing, the apparatus complied with all manufacturer's requirements, along with local, state, and federal laws, NFPA recommendations, and nothing else.

CONFLAGRATIONS

1804 Philip Mason Hose Wagon   1901 Pumper + Hose Cart

Pictures courtesy of C. Dewey

As populations increased, cities, towns, boroughs, and villages grew larger and larger. Office buildings in downtowns became increasingly higher. Conflagrations increased, as growth spread. Additional sophisticated equipment for firefighting was essential. Structures grew higher and higher, while the need for more space increased. Downtowns filled the sky with mammoth high-rise structures. Living in downtown hotels and apartments changed, as more and more office buildings emerged. The farmland was next to go, as 'bedroom' communities sprung up everywhere. Farmland was gobbled up. Then a new era of shopping that used to be only in the downtown stores, sprang up all over the countryside. Downtown buildings were becoming vacant; stores were closing their doors; they were losing customers to the neighborhood strip malls. The population had moved to the country. Something had to be done to the vacant structures, and to bring the people back to the cities. Alas, an innovation, "Condominiums." That was one way to save the downtowns. Also, rebuild the old stores into dining and shopping areas was another. With the 'bedroom' communities still growing, and downtown with an innovation, there was a demand for various types of additional equipment and apparatus to meet this new era. Manufacturers had to work very hard keeping up with the changing times. Equipment had to become larger and larger, stronger, and stronger, and higher and higher to reach these heights, and to fight the different conflagrations. Ways to merge multiple functions into a single apparatus were being developed. There will always be a need for new types of firefighting equipment to do a better firefighting job.

Pictures courtesy of San Francisco Fire Department

1607 - James Fort, Virginia –
The first permanent English Settlement in the United States
1619 - The first documented African slaves were brought into Jamestown
1676 - The Great Fire of Jamestown, Virginia - Burned down by Bacon's rebellious followers then later rebuilt

1776 and 1835 - Great Fires of New York -
1776 - Fire consumed over 400-500 structures

1835 - Fire consumed 50 acres and 500-700 structures

1788 and 1794 - Great Fires of New Orleans -
1788 - Fire consumed over 800 structures 1794 - Fire consumed over 200 structures

1849-50-51 - Great Fires of San Francisco -
There were seven large conflagrations in this period that consumed over three thousand dwellings

1850 and 1865 - Great Fires of Philadelphia -
1850 - Fire consumed many warehouses on the waterfront
1865 - A lake of fire from burning barrels of petroleum engulfed the whole downtown area

1866 - The Great Fire of Portland, Maine -
Fire consumed over 1800 structures

1871 - The Great Fire of Chicago -
Fire consumed over 500 structures - The possible cause was a 'cow' knocking over a lantern

1872 - The Great Fire of Boston -
Fire consumed over 700 buildings - Firefighters were hampered by a flu epidemic that immobilized the horses

1700 – 1870 – Other Great Fires
        1776 New York
        1845 Canton China
        1871 Peshtigo Wisconsin
        1871 Michigan

1600 – Present Day Arson Preset Fires – or Accidental

It is unknown just when the first Arson fire may have been set or recorded as an Arson fire. This has been a problem to the fire service and still to this day remains one of the high priority subjects. Many fires were set to collect large amounts of insurance moneys, some set by the property owners and there are some where the owner will pay someone to set the fire for them. Sometimes the arsonist is difficult to expose because they are part of the investigating group, or work for the fire departments or insurance companies. Disgruntled employees and upset persons, are also part of arsonists causing costly damage and sometimes injury and death. The largest is Forest Fires. They are the costliest of all fires and most difficult to extinguish. It may take considerable time to determine the cause of the fire, but most arsonists are eventually apprehended and penalized for what they have done.

Then there are the accidental fires caused by carless smokers, or cooking pots left burning on kitchen stoves that have caused excessive damage to apartment buildings and family homes. Backyard barbecues out of control or left to burn out have caused many home fires. Campers build small fires in the woods that sometimes get out of control by winds blowing embers into the woods causing grass or trees to ignite. These fires can increase is size and destroy homes and cities, especially in very dry and windy areas.

On the east coast of America, volunteer fire departments started as early as the 1600's. Volunteers fought fires for many decades with whatever equipment was available and wherever they could find a water supply.

In the 1600's to the 1700's, volunteer hook-and-ladder companies were formed to fight fires. Their functions were to pull down ceilings, and to ventilate the flames in building fires. Their ladders and poles were wood, and most were fifty feet or less in length. For centuries, fires were fought with bucket brigades, manual pumps, by volunteer ladder companies. In the early years of the new world, equipment was purchased from London, England, or other European cities, as they were the forerunners of fire equipment and had been fighting fires for centuries. The bucket brigade was the first and the only way to extinguish fires at first in the new world. Later, in 1653, the Boston Fire Department purchased their first engine pumper from London, England, to provide water-line pressure. Manual hand pumpers, with a narrow pipe to act as a nozzle, were used in the New York Fire Department until 1731, when they also opted to purchase two engine pumpers from London, England. The volunteers had a kind of pride working the manual pumps, and it was difficult to add the engine-driven pumpers into the system. Equipment needed to keep the fire department in full operation, and the maintenance necessary to maintain this equipment, was a very important subject not to be overlooked. Many varieties of pumpers and hose tenders were being developed everywhere.

1769 The Torrent – (6th in USA)
Picture courtesy of C. Dewey

As years became decades, larger cities were annexing outlying communities into their inner-city circle. Many of the fire department firefighters would maintain equipment at the firehouses. As time passed, the demand for maintenance increased, especially in the larger firehouses, so those departments began to hire full-time mechanics, sometimes a fireman would also be the mechanic, for the purpose of maintaining the equipment. Most mechanics had a dual workload. They also assisted with fires and emergencies. To service this equipment and keep the fire department's apparatus operating on a twenty-four-hour cycle, was the mechanic's most important task. Breakdown of equipment seemed to never cease. In smaller fire departments, the in-house mechanics had to move the apparatus around the firehouse bays, to service or do minor repair to the equipment. The mechanics always had a problem, when equipment had to leave the firehouse to respond to a fire call or emergency, as they tried not to affect the firefighters' routine. The task of just maintaining equipment was a chore, and the in-house periodic inspections were not in the curriculum.

Steam Engines:

Steam propulsion was developing in the early 1800's, in the U-K for rail locomotives. They used coal and wood fire to heat water for the steam. It wasn't until the 1812's when an American Matthew Murray developed the first commercially successful steam locomotive (Salamanca). They all had to use rails to propel forward to their destinations. In the early years of the nineteenth century, American manufacturers were beginning to supply fire departments with various types of firefighting equipment. In the 1850's, one of the first steam engines was manufactured by Lee and Lerned. When the tradition of using both the hand-drawn, manual-labor pumpers, along with the old engine pumpers began to phase out, the new steam engine commenced making its way into fire service.

1854 Wagon Approx.   1871 Horse Drawn
Hose Wagon

Pictures courtesy of C. Dewey

Annexing of the surrounding suburbs and towns, firefighting became more perplexed situation. This required both volunteer and paid firefighters. Firefighting had reached new heights. This would require various types of new and improved apparatus, turn out gear, ladders, aerials, and rescue equipment, to meet the demands of the generations to come.

The heights of the skyscrapers were steadily increasing, and multi-story tenement buildings were popping up everywhere. Skyscrapers were exceeding 10-20 stories, and extension ladders could not reach that height, as they were somewhere between fifty and seventy feet when fully extended and at full elevation.

In 1875, one of the first multi - extension ladder was tested. This aerial ladder had a height up to ninety-seven feet. This was a Scott-Udas aerial extension ladder that was manufactured in Canada. This extension ladder failed, when service tested, and three prominent men were killed. These longer extension ground ladders were then considered too dangerous for service. This forced the departments to continue using shorter extension ground ladders, and the firefighters were still forced to use scaling ladders for high-rise rescue, until an improved extension ground ladder became available. It was then believed that manual extension ground ladders over sixty feet would be too dangerous for fire service.

Manufacturers would begin making wooden extension ladders into truck-mounted aerials, with manual gear elevation and extension, and manual rotation. Many years passed before the wood aerial ladders were transformed into metal aerial ladders, using stainless, aluminum and hi-tensile steel as the basic material. Manual rotation, elevation, and extension functions were modified in the coming years to hydraulic power. This would be the invention for movement of the aerial ladder and platform operations. Wood 'aerial' ladders remained in service as well as wood 'ground' ladders for many years after the metal aerials were introduced into fire service.

"The bucket brigades now a memory from the past"

Picture courtesy of C. Dewey

FORMING A LADDER COMPANY

The town of Boston in 1631 formed their first volunteer fire department. It wasn't until 1737, when the first volunteer fire department was formed in New York City. It was said that the first organized Hook and Ladder Company was also in New York City in 1772. This company was known as the Mutual Hook & Ladder Company No. 1. History denotes that it was the forerunner of the present Ladder Company No.1 of the New York City Fire Department in 1952. On June 16, 1784, company members met and reorganized. They re-took the name of "Mutual." There appeared to be no description of the type of ladders and/or equipment used during that period. Makeshift ladders have been in existence since before BC. As home dwellings grew, different versions of ladders were made to gain access to difficult areas. Ground ladders with hooks were used in this country as early as the late 1690's. Prior to the 1800's, departments were experimenting with extension ladders. The records are unclear as to when extension ladders were first put into service and trying to reach the top of any high-rise building was an ongoing problem that needed to be solved. One of the first ladder trucks in this country was driven by oxen, and they broke no speed records getting to a fire.

Picture courtesy of Fire Engineering

The "Mutual," on November 8, 1832, purchased a horse for $88.00; and then in April 1833, the horse was sold with no reason stated. In 1836, a committee was appointed to dispose of a horse that was considered unfit for duty. Possibly, the reason was firemen were so used to pulling the equipment to the fires by themselves, they didn't understand how useful horsepower could be.

In 1848, the "Mutual," received a new fire truck "of which they felt very proud," considering it was the 'handsomest' truck in New York City. In 1849, it was said that the brass work on the truck was silver-plated.

Folding Ladder Truck of Hook and Ladder Company No. I, New York, in 1855 .

Picture courtesy of Fire Engineering

In 1855, the "Mutual," now has a Folding Ladder Truck. What was meant by "folding," was unclear, but possibly, they were referring to the tiller-steering arrangement, whereby the apparatus could be "folded" or "turned" in a narrow arc.

It was said that the "Mutual," never lost their organization status for a single day. During the existence of the **Old Volunteer** fire department, and on the founding of the **New Paid** fire department, on September 8, 1865, using the same location, the same truck, and the same distinguishing red cap fronts as the Mutual Hook & Ladder No. 1, the "Hook & Ladder Company No. 1," was officially created. It could be stated that the Hook & Ladder Company No. 1, was in continuous operation since June 16, 1784.

In 1880, the first Aerial Ladder purchased in New York City was manufactured by Daniel Hayes. It was horse-drawn, with a tiller driver seat mounted over the rear axle. This ladder had to be manually operated. After significant trial and error of multiple-section extension ladders, it was finally established that wood extension ladders should not exceed a maximum of seventy feet in length.

Picture courtesy of Fire Engineering

FIRE EQUIPMENT MANUFACTURERS:

To assist in extinguishing fires and for rescue, firefighters need to know a little information about the manufacturers who furnish equipment and put firefighting into a professional business.

Many years have passed since the forming of fire departments in the New World. Equipment was either purchased from foreign countries or made makeshift to fight fires and rescue. There were very few records kept before the 1800's, on firefighting equipment. The early nineteenth century started the boom of fire equipment manufacturers in America. Some of the original manufacturers continue to do business and have survived all economic strife. At the beginning, many of the manufacturers were in business, for only a short period of time, or merged with larger companies. With the arrival of the twentieth century, fire equipment manufacturers of many categories were everywhere. The following list may not contain all the manufacturers who contributed to this great field; but somehow, they all know they played a part, and have contributed to the future of fire department safety. This list will begin with one of the earliest manufacturers of fire apparatus made in America and continue through the current manufacturers of the twenty-first century.

American LaFrance - 1832 - Present

History drops back to 1832, when the first fire equipment company formed, was to become American LaFrance. They started manufacturing various types of fire apparatus. Then, founded by Truckson LaFrance, in 1873, the company was to become LaFrance Mfg. Company. In 1886 they would merge with Daniel Hayes, in the manufacturer of 290, 75-foot horse drawn tiller wood aerial ladders. The aerial ladders were made from wood until the early 1900's, when steel aerial ladders were introduced. In 1903, American LaFrance Engine Company was formed. This company also operated a manufacturing business in Toronto, Canada, called LaFrance-Foamite. In the 1960's, American LaFrance developed the Articulating Aerochief seventy-, eighty-, and ninety-foot aerial platforms. In 1971, American LaFrance purchased Snorkel Manufacturing, another articulating aerial device. At that time, American LaFrance was the top manufacturing company of fire apparatus, with operating facilities in New York, Georgia, Missouri, California, and Canada, who gave full service to their customers. Present day, American LaFrance, is still in production and has a complete line of equipment--Pumpers, Telesqurts, Aerial Ladders, Ladder Platforms, and other associated equipment.

Pictures courtesy of American LaFrance

Daniel Hayes - 1868

Daniel Hayes, another pioneer of fire apparatus, was working for the San Francisco Fire Department in the mid 1860's. His position was Superintendent of steam engines. In 1868, Daniel developed one of the first aerial extension ladder apparatus in America. He developed and built several pumpers and over two hundred ninety aerial ladders. San Francisco used the Hayes aerial extension ladder until the late 1950's. Daniel Hayes represented American La France of Elmira, New York.

Seagrave - 1881 - Present

In 1881, Seagrave was manufacturing a full line of fire apparatus in Detroit, Michigan. In 1891, they moved their operation to Columbus, Ohio. Canada was another important division for Seagrave Manufacturing. From 1908 to 1936, Bickle-Seagrave was the Canada division. Bickle then worked under the Bickle-Seagrave banner until 1956, when King-Seagrave was formed. It wasn't until 1963, when Seagrave was purchased by FWD Corporation, and they set up corporate headquarters and manufacturing in Clintonville, Wisconsin, where they continue to manufacture fire apparatus today. King-Seagrave worked together until 1973, when FWD Corporation opted not to continue relations with King. Seagrave's Canadian representation is presently in the city of Carlton Place, Ontario.

Seagrave aerial ladders were made from wood until the mid-1900's, when they changed over to lattice-type steel telescopic ladders. Seagrave manufactured a light duty and a now has a heavy-duty steel aerial ladder, in addition to pumpers, and a variety of universal firefighting equipment. Seagrave acquired the manufacturing rights to the Baker-Mack seventy-five foot and ninety-five- foot Aerial platforms, from Baker Manufacturing in Richmond, Virginia. In 2003, Seagrave became a flagship of ELB Capital Management. Presently, Seagrave is in production with a versatile line of pumpers, aerial ladders, aerial ladder platforms, Aerial platforms, and other associated firefighting equipment.

Pictures courtesy of Seagrave Fire Apparatus, LLC

1937 Seagrave Picture courtesy of C. Dewey

Peter Pirsch and Sons - 1890 - 1989

In the 1890's, Peter Pirsch and Sons were manufacturing a line of truss extension ladder trucks, which were hand-pulled and horse-driven. They were incorporated in Kenosha, Wisconsin. Their aerial ladders were manufactured from wood, and later converted to riveted aluminum. From the 1920's through the 1980's, were their most productive years. The great-grandson became President of the company in the late 1970's. Peter Pirsch and Sons were in business for almost one hundred years, when they closed their doors. Peter Pirsch and Sons manufactured their own style of solid aluminum ground ladders with single and multiple sections. Peter Pirsch and Sons were one of the real pioneers of the fire apparatus industry.

Pictures courtesy of C. Dewey

Picture courtesy
of C, Dewey

Picture courtesy of C. Dewey

Sutphen - 1890 - Present

With over one hundred sixteen years of being a sole family run business in Columbus, Ohio, the Sutphen family started manufacturing their first piece of fire apparatus in one of their employee's garages. For decades, they were custom manufacturers of various types of fire apparatus; they were very successful; but they were cramped for more manufacturing space. In 1958, working out of only three shop bays doing custom manufacturing, they increased their facility to five bays. In 1964, the company relocated to their present plant in Amlin, Ohio, a suburb of Columbus, Ohio. Today, Sutphen has five manufacturing factories, and a complete line of fire apparatus, to meet the needs of any fire department. Their aerial ladders and aerial platforms are constructed from aluminum and fastened together with aircraft-type huck-bolts. The truck bodies are made from stainless steel and/or aluminum.

Hahn Fire Apparatus - 1898 - 1989

After making horse drawn carriages for many years, in Leesport, Pennsylvania, Hahn started making fire trucks in the early 1900's. Later in the 1920's, Hahn Motors started building a custom chassis, in Hamburg, Pennsylvania, for commercial trucks and fire apparatus. For decades, they supplied the chassis for various types of apparatus. In the 1970's, Hahn was manufacturing a telescopic 106-foot aerial ladder, called the "Fire Spire." In 1989, Hahn had to close their doors to the manufacturing of fire apparatus. In 1991, a special retrofit kit was installed on all existing Hahn one hundred six foot "Fire Spire" ladders as a safety precaution.

Pictures courtesy of C. Dewey

Mack Trucks, Inc., Fire Apparatus - 1900 - 1990

Mack started their first operation in 1900, with the manufacture of the Mack Bus. It was not until 1911, when they started manufacturing fire trucks in Allentown, Pennsylvania. Mack discontinued manufacturing fire apparatus in 1990.

Pictures courtesy of Mack Truck

My first contact with Mack and their fire department equipment was when I assisted with the new design of the seventy-five-foot Mack Aerialscope, in the early 1960's, in Denver, Colorado, while I was working for Truck Equipment Company, "Truco;" a Denver-based utility equipment manufacturing company. Fire Apparatus was to be "Truco's" next adventure in manufacturing. This was to be the first 'Telescopic' aerial platform, made for fire rescue, with an elevated nozzle service. Mack was searching for a fire service aerial platform to mount on their chassis. They have one of the best chassis structures for mounting this type of equipment, and multiple manufacturers of pumpers use their chassis.

Mack first was considering an 'Articulating' aerial platform, called the "Snorkle," which was manufactured from 1958, by Pitman Manufacturing. Mack decided not to go with the articulating type of boom structure, and instead went with the multiple extension-type boom sections. My duties were to assist in the original design of the seventy-five-foot Mack Aerialcope and develop an intricate electro-hydraulic system for operation of the boom movements, and to assist in quality control inspections for the aerial division.

Manufacturing of the seventy-five-foot Mack Aerialscope, changed hands three times. "Truco" continued through the early 1960's, but later was sold to a local company, Eaton Metal Products of Denver, Colorado, a family-owned farm tank manufacturing company. A few years later, the Mack Aerialscope portion Eaton Metal's business was sold to Baker Engineering of Richmond, Virginia. Baker was a utility equipment product supplier and manufacturer. Baker started manufacturing the seventy-five-foot Mack Aerialscope, and later developed a ninety-five-foot version of the Scope still mounted on a Mack chassis. It was after I started PTS, Professional Testing Systems in the early 1980's, that Baker Engineering gave me the privilege of performing most of their final inspections for acceptance for various fire departments, who had purchased the seventy-five or ninety-five foot "Scope" mounted on the Mack chassis. For many years, Baker exhibited both 'Scopes,' at Conventions; and product sales were excellent. Then, as the economy turned, Baker fell on hard times, and the Mack chassis mounted Aerialscopes, were sold to Seagrave Manufacturing Company, where they are still under manufacturer.

Thibault / Pierreville - 1908 - 1990

Professional Testing System was the first selected U.S. Inspection and fire apparatus Testing Company, used by a Canadian Manufacturer, for "Third Party" Inspections and final acceptance of fire apparatus. These acceptances of apparatus were for fire departments in both the U.S. and Canada.

In 1908, a Canadian family of seven brothers named Thibault started manufacturing hand-pumpers in Saint-Roberts, Quebec. A few years later, they moved their operation to Sorel, Quebec, and manufactured horse-drawn apparatus, with some pumpers mounted on sleighs. In 1918, they built their first motorized fire apparatus. In 1938, the operation moved to a small town called Pierreville, Quebec, next to the St. Lawrence River, and close to a nearby Indian Reservation. In 1960, they introduced their first steel telescopic aerial ladder. In 1963, their advertising showed the strength of the aerial, by hanging a Volkswagen Beetle at the tip of the ladder base section. Throughout the years, there were many disputes between the seven Thibault brothers. They split up and later formed two different fire apparatus manufacturing companies; one called Pierreville Fire Trucks, and the other called Thibault Fire Trucks, which was the original manufacturing company. Pierreville Fire Trucks was later forced out of business in 1984; and at that time, Thibault Fire Trucks was renamed Camions Pierre Thibault, Inc. Despite the hundreds of pieces of fire apparatus manufactured in the 1980's, Camions Pierre Thibault, Inc. was forced to declared bankruptcy in 1990. In 1991, their assets were purchased by a group of investors, and Nova Quintech Corporation was formed. Nova Quintech Corporation discontinued manufacturing pumpers and concentrated only on aerial ladders and aerial platforms. In 1997, Nova Quintech Corporation sold their manufacturing rights to Pierce Manufacturing, thus ending a saga of the seven 'unique' Canadian brothers.

The name "Thibault," was always pronounced "Tibo," as per the late Marian Thibault, the oldest brother, and the strongest entrepreneur.

Pierreville 100 Ft. Steel Aerial Ladder Picture courtesy of Pierce

Pierce - 1913/1940 - Present

In 1913, Humphrey and Dudley Pierce started manufacturing after-market truck and bus bodies in Appleton, Wisconsin. It was in 1940, that Pierce was asked to build a fire truck body on a commercial chassis. During the next forty years, many changes were made to improve the quality of their products. In the 1980's, Pierce's heavy- duty steel aerial ladders and aerial ladder platforms became one of the top lines of the industry. "Bronto

Skylift," is one of Pierce's largest and tallest articulating aerial platform products. In 1991, Pierce introduced their new one-hundred-foot aerial ladder platform with the "Pierce Micro." In 1996, Pierce was acquired by Oshkosh Corporation. In 1997, Pierce purchased the manufacturing rights of Nova Quintech, a Canadian manufacturer of aerial fire truck products. In 2008, Pierce introduced their new seventy-five-foot aluminum ladder, an expansion of their "PUC" line. Pierce now has the most unique complete product line of fire apparatus in the industry. In the 1980's, at the beginning of "Third Party" Inspections, PTS had the opportunity to work with Pierce Manufacturing for final acceptance of new equipment sales.

Pictures courtesy of Pierce Manufacturing

Maxim Motors - 1914 - 1989

Maxim Motors was the pride of Middleborough, Massachusetts, and the Cape Cod area, along with sales throughout the U.S. and Canada. Their first fire truck was built in 1914 and started a long life for Maxim in the fire truck manufacturing industry. It wasn't until 1949, when the first Maxim seventy-five-foot aerial ladder was delivered to Falmouth Fire Department and served the Cape Cod area until 1968; at which time, the Hyannis Fire Department bought their first new aerial ladder. One of the unique features of the Maxim aerial ladders is they are all made from stainless steel. As years passed, the aerials became longer, and the quality of their pumpers greatly improved. Maxim officially closed their doors in 1989, but their memories will linger in the Cape Cod area forever. Middleboro Fire Apparatus, a repair facility in Massachusetts, continued to represent Maxim equipment and work the Maxim line of products for several years to follow.

In 2009, Greenwood Emergency Vehicles of North Attleboro, Massachusetts, announced the reemergence of the Maxim brand of products.

Pictures courtesy of Maxim / Greenwood Fire Apparatus

Oshkosh Corporation - 1917 - Present

In 1917, Oshkosh Truck Corporation was founded and was the first to build four-wheel drive trucks. In 1968, they introduced the MB-5 aircraft and firefighting vehicle for the U.S. Navy. In 1977, another more powerful aircraft firefighting and rescue vehicle, called the P-15 ARFF, was developed. In 1996, Oshkosh Corporation acquired Pierce Manufacturing and became the world's leading fire truck manufacturer with a vast number of versatile firefighting apparatus. In 2001, Oshkosh acquired the "Snozzle" aerial device and developed the Striker ARFF, which replaced the "Snozzle." Oshkosh continues to be one of the top manufacturers of fire department and airport firefighting equipment.

Kovatch/KME - 1946 - Present

Kovatch Mobile Equipment Corporation started manufacturing custom specialty vehicles in a two-car garage in Nesquehoning, Pennsylvania, in 1946. For the next several decades, the business grew; and presently, there are eleven manufacturing plants at that facility. In 1983, they began manufacturing various types of fire apparatus. In 1985, they changed their name to KME Fire Apparatus. "KME" took over the manufacturing of the Grumman "Aerialcat," a heavy-duty steel aerial ladder platform. Today, "KME" has a versatile class of emergency vehicles and aerials to supply any fire department's needs. One-of-a-kind manufacturing challenges are "KME's" specialty, with a 'can do' attitude.

Grove Manufacturing - 1947 - 1980

John L. Grove was a pioneer in hydraulic cranes and manlifts. He built his first extensible hydraulic chain-driven aerial ladder for fire service in the late 1950's. This design was of higher capacity than the pre-1980 manufactured aerial ladders. In 1967, Grove Manufacturing was bought out by Walter Kidde Company. In the 1980's, the Grove ladder was the inspiration for new-style manufacturing of aerial apparatus, and the start of a new era for heavy-duty telescopic ladder trucks.

Pitman Manufacturing - 1950 - 1990

Pitman has been a leader in product innovation since 1950, developing a variety of equipment for the utility and construction industries. In 1958, the "Snorkel" was developed for fire rescue and for elevated water nozzle usage. The articulating aerial platform reached heights from fifty-five to eighty-five feet. The Chicago Fire Department was one of the first to use this articulating type of aerial platform with a water nozzle. Pitman Manufacturing no longer manufactures the "Snorkel," but Altec NUECO Corporation distributes both Altec NUECO and Pitman lines of utility equipment.

1967 - 75 Ft. Pitman "Snorkel" Pictures courtesy of Reliance Fire Museum

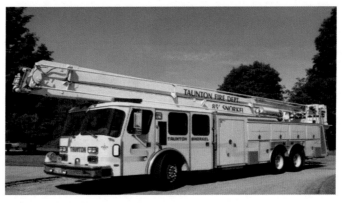

85 Ft. Pitman "Snorkel" Picture courtesy of Taunton Fire Department

Smeal Fire Apparatus - 1955 - Present

Starting as a small welding shop in Snyder, Nebraska, in 1955, the founder, Donald Smeal, a volunteer fireman, kept together what is today a prominent fire apparatus manufacturing corporation in Nebraska. In 1964, Donald designed his first forty-two-foot aerial ladder. Today, there have been more than four thousand custom aerial ladders, aerial platforms, and pumpers manufactured to suit all the needs of many fire departments throughout the world.

Pictures courtesy of Smeal Fire Apparatus

Mobile Aerial Towers Hi-Ranger - 1959 - 1992

Mobile Aerial Towers was primarily in the manufacture of an articulating fiberglass boom truck, designed for Utility Departments, for power line work, and for the tree trimming industry. In the early 1960's, the demand to extinguish fires, and rescue stranded people, in areas where ground ladders could not reach, prompted the manufacture of the Hi-Ranger Aerial Tower. One of the first Towers manufactured was a seventy-five-foot articulating platform with open-lattice booms, and a pre piped water line with a nozzle at the front of the platform. Mobile Aerial Towers continued manufacturing this standard equipment until Thirty-three years later, when Telelect Corporation purchased the manufacturing rights to Hi-Ranger, and now Telelect Corporation is a division of Terex Corporation, a fortune five hundred company.

Calavar Corporation - 1970 - 1980

In the early 1970's, a new bird was born in Santa Fe Springs, Ca. The big bird was called the Condor Firebird. Many Condors were manufactured for the construction industry. A multi-action articulating, and telescopic firefighting platform, with heights from seventy-five-feet to one hundred and fifty feet, was offered. The Condor was short lived, and construction ceased in 1980. One of the fire departments that made good use of the Firebird was the City of Richmond, Virginia, and they probably had the only one-hundred-foot Firebird used for firefighting.

Pictures courtesy of John E. Hinant, Richmond, Virginia, Fire Department 'Retired'

E-ONE Corporation - 1974 - Present

In 1974, E-ONE was founded in Ocala, Florida. Their products are constructed mostly of extruded aluminum. E-ONE has a vast versatile product line that exceeds most other manufacturers. The aluminum aerial ladder lines have models from seventy-five to one hundred and thirty-seven feet in length. Airport crash rescue trucks, along with many models of pumpers, tankers, and command units, are but a few of the E-ONE product lines. E-ONE also has a full line of stainless-steel apparatus products, which are manufactured at their Hamburg, New York, facility. In 1993, E-ONE purchased Superior Equipment Company; and in 2004, they purchased Saulsbury Equipment Company. Both have fine lines of fire equipment.

E-ONE is a large operation, manufacturer, and distributor of a superior line of aluminum firefighting products. E-ONE has the industry's leading aerial safety record – Zero tip-overs and structural failures. E-ONE builds one of the strongest chassis on the market, making them a sole-source manufacturer.

PTS has had the privilege of inspecting and accepting new apparatus for fire departments at the E-ONE facility for many years.

Pictures courtesy of E-ONE

Steeldraulics Corporation - 1985 - 1987

One of the first heavy-duty aerial chain-driven extension ladders, built for fire service, was manufactured by Grove Crane. Grove was going out of the fire service business; therefore, a new company was formed, called Steeldraulics Corporation, located in Pennsylvania. The company developed the first heavy-duty aerial ladder for firefighting. This was the start of heavy-duty type aerials. Their manufacturing was short lived, and the product was sold to two new manufacturing developers of fire apparatus. Grumman manufacturing located in Roanoke Virginia, and Ladder Towers, Inc., located in Pennsylvania. Also, this was the beginning of a new era in the manufacture of heavy-duty steel and aluminum aerial ladders for firefighting and the ladders continue to be improved. Grumman did not last in the aerial fire service long, and eventually sold the manufacturing rights.

Ladder Towers, Inc. - 1980's - Present

Ladder Towers, Inc. was one of the first manufacturers to build a heavy-duty aerial ladder and telescopic aerial platform, with a pre-piped waterline and horizontal reach capacity. At first, the aerials were chain and cable extended. Later, chain and sprockets were replaced with hydraulic extension cylinders, and cables. LTI manufactured a mixed variety of fire apparatus--airport crash rescue trucks, and pumpers with various capabilities. LTI also manufactured one of the most unique tractor-drawn aerial ladders, with replaceable compartments in the body. When a compartment is damaged, simply unfasten, remove, and replace. LTI was purchased by Simon Manufacturing Company of the United Kingdom. In 1987, Simon was manufacturing a Super Snorkel, SS600, which could reach heights up to two hundred and two feet (61.5 meters). At that time, Simon purchased various utility manufacturing corporations in the United States. Later, Simon LTI went through another change of owners, and finally was purchased by American La France, who continues to manufacture this elite product line today.

Picture courtesy of American LaFrance

Snozzle - 1980 - Present

This is a very special unit, and most firemen do not even know its history. In 1980, while I was inspecting fire apparatus on the East Coast, I originated the idea for the Snozzle Aerial Lift. I gave this idea to a long-time friend/partner, who was an engineer living in California. Together, we developed this apparatus for both utility and fire service. We decided not to pursue the utility side, but only concentrate on its development for fire service. I installed the first firefighting Snozzle apparatus at the Third District Fire Department in River Ridge, Louisiana, in 1981. After installing this unit, I was working nearly every day as a Third-Party Inspector, and I lost contact with my engineer/partner for a short period of time. The next time I heard from my friend, he had patented the Snozzle product and quickly sold it to a Texas company. In 2001, Oshkosh Corporation purchased the manufacturing rights to the Snozzle, and it is now called the Striker ARFF. Oshkosh Corporation has a superior line of firefighting apparatus, including pumpers, airport crash-rescue trucks, and four-wheel drive vehicles. The Snozzle will blend in very well for airport safety.

SUMMARY: Aerial ladders, aerial ladder platforms, aerial platforms, quints, pumpers, rescue, and airport specialty equipment of today, are safer to use, stronger, and designed with many more functions and safety features included for rescue and extinguishing fires, than the fire equipment of yesteryear. Pre-piped waterways and automatic nozzles, with larger flow capacities, make it less complicated and quicker to extinguish fires. An enormous change has evolved from the beginning, 'bucket brigade' to our present day, 'firefighting apparatus.'

"THIRD PARTY" INSPECTIONS:

Aluminum Aerials:

There are manufacturers of aluminum aerial ladders and aerial ladder platforms that use either rivets, Houck fasteners, or simply weld the extruded structures in place. There are different ways to inspect these units for wear or rivet looseness, according to how the rails are assembled. Visual inspections can identify problems about ninety-nine percent of the time. The other one percent is to use an NDT procedure that will check for internal flaws, and indications of cracks. Ultrasonic inspection will show if there are any internal cracks in the rivets and fasteners. A portable hardness tester placed in many areas against the aluminum structure will indicate any loss of tensile strength or weakness from heat. In the field, there are a few "Third Party" Inspection Companies using acoustic emission testing for aluminum aerials. This special type of inspection uses a series of transducers placed throughout the ladder rails, or along the boom section structure. When the ladder sections are load tested or stressed to a point where the transducers pick up stress noise, it is noted which transducer received that noise on the readout. This does not always indicate structural damage, only an area that flexes or makes noise. Riveted aluminum aerials make a lot of noise just extending and retracting. On aluminum weld structures, dye penetrates can be used to verify structural damage where a visual indication of a crack is noticed. In the early 1980's, one of the larger manufacturers of welded aluminum aerials had problems with "Third Party" Inspection Companies placing many units out of service. The manufacturers welding procedure at that time would leave many welds with small crater crack indications at the end of the weld. These indications were caused when the weld cools, and this non-structural weld puddle shows up with a surface indication. This problem caused much trouble for the manufacturer, when the Third-Party inspector in the field, was removing apparatus from service. The manufacturer quickly corrected this minor welding situation, by eliminating further crater cracks. An information flyer was then sent to all "Third Party" Inspection Companies and all fire departments using this type of apparatus, explaining their procedure on this subject; and the process for correcting this problem with units in the field. Some Fire Department Chiefs were convinced steel aerials would outperform aluminum at structure fires since aluminum has a lower melting point compared to steel. Throughout the years, aluminum aerials have proven to sustain all the elements with endurance and have very low maintenance. The hardest part of convincing a fire department, a manufacturer, a municipality, or just anyone who needs fire apparatus inspections, is to prove the experience and qualifications of the "Third Party" Inspection Company, and their inspector. In the last few decades, the qualifications and expertise of "Third Party" Inspection Companies was not of the highest quality. Presently, the way the economy is, money and "low bid," usually wins. It doesn't seem to be important about the amount paid for the equipment, or how thorough the inspection would be, but only how fast and inexpensive the department can get the Certificate of Compliance to present to ISO, or any other officiating agency, that would prohibit the use of the equipment, without a certification. There are many large fire departments that will bundle a group of outlining departments just to get a low bid. The problem is that the low bid inspector would be cutting down on the time required for an individual apparatus inspection and would double up on inspections. Who will eventually come out on the short end of this deal? After a period, the equipment areas that did not have proper inspections will start to deteriorate, and the cost of repairs will greatly exceed the cost for one 'complete' professional apparatus inspection per day.

Third Party Inspector:

How qualified and knowledgeable of your equipment is the "Third Party" Inspector your department has hired? "Third Party" Inspectors vary from retired firemen looking for extra work, fire mechanics looking for side jobs, inspectors just graduated from high school, utility linemen with some apparatus experience, NDT technicians with no apparatus experience, crane, and derrick inspectors, and to whomever. When you hire a fire apparatus inspector, one thing you need to know, there are no schools or certifying agencies to train fire apparatus inspectors. What governmental accrediting agency makes them qualified for fire apparatus inspections? "None, that I recall." So, "Who tests the testers?" The Department of Labor accredits engineers, and inspectors with experience in the design of cranes and derricks, with years of experience in servicing and inspecting same. "A Department of Labor Accreditation" was all that "Third Party" Inspection Companies had to use for some type of inspection qualification. As I stated before, fire department aerials are like cranes and derricks, but they just don't have water pipes. California is the only state in the union that has its own "Third Party" Inspection Company accreditations. There are special qualifications that must be met, and tests to be taken, to apply for accreditations in the inspection field. Inspectors that are trained by accredited engineers or agencies must work under their scrutiny, and they must approve all reports. It takes years of on-the-job training to familiarize oneself with all the makes, models, and critical areas on fire apparatus to be inspected. A fire apparatus inspector must start somewhere; and through many years of inspecting various types of apparatus, the inspector should have learned something. If the inspector does a quick inspection and never finds any critical problem areas, I hope for fewer apparatus failures.

The "Third Party" Inspection Company, where I was first accredited as a crane and derrick inspector, was in the state of California. There were partners in that company who branched out to another state, and continued crane, derrick, and fire apparatus inspections. When the California based company closed its doors, the branch office had to be re-accredited by the Department of Labor since it was not located in California. I was given the opportunity to stay in California and continue inspecting cranes and derricks under my "Third Party" Inspection Company name. Fire apparatus has always been my preference, but at that time, seventy-five percent of fire apparatus was located on the East Coast, so I relocated to Virginia. In 1978, PTS became a "Third Party" Inspection Company, incorporated in Virginia Beach, Virginia, located directly between Miami and Boston. Then, it was time to become accredited by the U.S. Department of Labor, and prove to fire departments, that even though there are no certifying agencies for fire apparatus inspectors, PTS had credentials for designing and inspecting a variety of apparatus.

In this world of free enterprise, sometimes it is not what you know, but what you must pay for.

"There is a story behind this." Paperwork for this accreditation was filed with the Department of Labor, and an appointment to see the director who issues accreditation certificates was set. Monday morning, I took my wife on a trip to Washington D.C. It was a beautiful day, the sun was out, and we expected to return home to Virginia Beach early in the evening. This would be the beginning of many long years of working with fire departments as PTS. We arrived promptly at 9:00 A.M, (I will call him Bob) he hadn't arrived yet. I glanced over to his desk and saw my accreditation folder on the top of a stack of other folders, and thought this is a good sign, and he is ready for me. At 10:00 A.M., Bob arrived at the office. First, he said that over the weekend, he went to this special mountain resort, fished, and had a great time. Bob then elaborated for the rest of the morning about all the special events he attended and gifts he received. Soon, it was 1:00 P.M. and Bob suggested we go to lunch at this special restaurant he enjoys. After I treated Bob to lunch, we returned to his office. Bob stayed very quiet and just looked at us as if waiting for me to say something. The folder was still on top of his desk; and at the end of the day, it remained unopened. It was getting late when I finally stood up and asked Bob, in a sarcastic way, how he got his job with the Department of Labor, that I was not going to give him one red penny for my accreditation, and that only my qualifications were all he would receive. Bob replied that his brother works downstairs, and he got him his job at the Department of Labor. We drove back to Virginia Beach empty handed. The next morning, by accident, I met with our company attorney, and told him about my experience at the Capitol. He told me he was going to D.C. next week. The following week, I got a call from Bob, at the Department of Labor. He asked me if I was trying to get his job. I explained to Bob, that I went there for my accreditation, and I did not want his job. Apparently, my attorney discussed this issue with a Senior Senator, and the Senator had a discussion with Bob. Two weeks later the accreditation was in the mail, and we were in the fire apparatus inspection business. Now, I can tell why I told this story. It was not long after I received my accreditation from the Department of Labor when a letter arrived through the mail from a Congressional Investigating Committee, that Bob had been terminated, and he was no longer with the Department of Labor. The Committee was now investigating eight hundred fraudulent accreditations given to "Third Party" Inspection Companies and individuals. What this meant was there are eight hundred testing companies, working in the U.S. under false pretense. PTS was investigated, and continued its accreditation through the mid 1990's.

There are "Third Party" Inspection Companies that hire overqualified or highly certified professional to write qualifications for their company, just to get accreditation for special circumstances. Once the company has the accreditation they need, they fire or layoff the expensive certified professional who wrote the qualifications and replace them with a professional with fewer qualifications to work for them under their new accreditation. Now you have an accredited company with no real experienced management personnel. They must depend on the professional to run the operation, hire inspectors, and hopefully have enough experience to supervise these inspectors so no one gets hurt and no equipment is damaged. Then, there are corporations that simply purchase the inspection rights from another company, and they may or may not have experienced personnel, but they have a license to inspect. Cities, municipalities, and fire departments that hire these companies never know their circumstances. It is a bad situation, but it really exists. Another reason the "Third Party" Inspection Company should have accreditation, is the difficulty to obtain liability insurance, and that becomes a legal issue. Unless the "Third Party" Inspection Company is financially self-insured, there are very few insurance companies that will take a risk, because the insurance premium is like errors and omissions. There have been inspection companies who received an insurance policy for a year period, then were canceled after the expiration date, not able to renew their policy, or couldn't find an insurance company to re-insure them the next year. To bid a future contract, some companies have taken their old, expired insurance contract, and simply changed the expiration date, and used the old insurance contract for bidding. This scenario has happened to a department that had multiple pieces of apparatus, and they required a minimum liability insurance policy. Even though "Third Party" Inspection Companies only give Certificates of Compliance showing that "the apparatus, at the time of the inspection, passed all applicable requirements at that time only, and there are no guarantees, that tomorrow, the apparatus won't fail in some way or other, unless the "Third Party" Inspection Company is financially stable and self-insured, most insurance companies will not insure "Third Party" Inspection Companies.

The way the economy is, presents a real problem that fire departments need to address. It is great if an inspection company can do multiple units daily due to the high costs of being in business. Still, with this situation, these departments will not get full inspection benefits. It has truly become a catch twenty-two for many "Third Party" Inspection Companies, and for fire departments.

I remember bidding a contract for multiple fire departments in Alaska one year, where there were no roads to access many of the cities, and inspectors had to fly to most cities. There were only three "Third Party" Inspection Companies in business at that time. Only two bids arrived, and PTS was the higher bid. After the awarding of the bid, this "Third Party" Inspection Company was on location in Alaska, and they had to ask the fire departments for more money to finish the contract, as they did not foresee the transportation problems when bidding the contract. The fire departments did allow some additional expenses, but never used that company again. It was said that a lot of ground ladders were also destroyed. This happened to be the company that purchased the inspection rights to fire truck inspections in California. Later, they sold the company inspection program to a nonprofit organization. Many red flags have been waved, and this was one way to get your attention. It **does not** always mean the company with the big guns has the best inspectors, or always does the best job.

BUYING AND SELLING APPARATUS:

USED AERIAL APPARATUS 'CAUTION'

Replacements of aerial apparatus after years of use and exercise, and the new design and upgrading of equipment promoted one of the next industries to emerge. This was a new channel for retired or worn-out equipment, and was appropriately called, "Used Fire Apparatus Sales." Individuals and small companies who sell 'used' fire apparatus would go to fire department salvage yards, bid on used equipment on the sale line. They would find fire departments that had just purchased a new aerial and were selling their old units. Usually, if the apparatus was in fair shape, they could buy it at a very reasonable price. The next step was to have a mechanic or helper, inspect the apparatus, and repair or replace all visual worn items. Then, they would clean and repaint the apparatus, so it would cosmetically look like new. When there was a potential buyer for the used apparatus, the company would call for a "Third Party" Inspection Company, to issue a Certificate of Compliance. PTS was called many times, by various used apparatus sales companies all over the U.S., to perform inspections on refurbished aerial apparatus. While inspecting the apparatus, we found many critical areas on the equipment had been overlooked and needed much more than cosmetic repair, so the equipment was not certifiable, and most was too unsafe to put into service. PTS declined inspecting apparatus for any of those used fire equipment companies, after we saw what was happening by other certifying companies.

Shouldn't new or used aerials be in good shape, and safe to use, if the maintenance had been performed correctly? Many times, the mechanic, or inspectors of these Used Fire Apparatus Sales Companies, do not know where the critical areas are or have overlooked them and approved the unit for sale, not knowing there were problems existing. Sometimes, a Third Party Inspection Company issued a Certificate of Compliance for a quick sale. When a fire department buys a used aerial apparatus from another fire department, or from a used equipment company in an "as is" condition, they should verify the apparatus has been properly maintained and tested prior to selling, and that certifying documents and reports are presented, showing the condition of the apparatus. Now, the aerial should be O.K. to purchase, but "Not all the time." The lack of training for maintenance personnel, in-house inspectors, "Third Party" Inspectors who have limited experience or knowledge of the equipment, can still leave the apparatus in jeopardy. After acceptance of an apparatus, the in-house mechanics and fire department inspectors who only know the basic operations of the newly delivered unit and have not been shown the critical stressed areas to inspect when servicing or inspecting the unit, should not be held responsible for any failure.

"Third Party" Testing Companies could have an inspector with limited knowledge of critical areas and with very little OJT experience in the field, and that does not help the process. There are fire departments that have used the same "Third Party" Inspection Company and same inspector, year after year. There is no guarantee these inspectors gave the department an honest or professional inspection. The fire department Chief or mechanic must assume a "Third Party" Inspector performing the inspection of their equipment knows what he/she is doing, so he would not have to wait and watch the inspection process. A short time after PTS closed its doors, I notified one of the older inspection companies around and ask them to let me check how they were performing inspections for fire departments. If I could not find problems with their procedure and the inspector did not call their office and say they were not inspecting properly, I would work for free. The challenge was accepted. I was to assist two 8-year 'experienced' inspectors employed by this "Third Party" Inspection Company, to re-inspect aerials they were already testing in two different states for many years. Any information found was to be used for training and to pinpoint any deficiencies the inspectors may have missed. There was a variety of fire apparatus in both small volunteer and large city departments. The very first day of inspecting pre-inspected aerials, the inspector called the office. After two months of assisting the two inspectors, many of the aerials had to be removed from service due to major problems that had been overlooked by both inspectors, who had checked the same apparatus year after year, and had not found these critical deficiencies. Many of the fire departments were very lucky, as they never had to set up their aerials, in an emergency, where the damage could increase, resulting in possible failure. I found many overlooked deficiencies while re-inspecting these aerials. Before taking any apparatus out of service I made sure the Chief or acting officer was visually aware of any deficiencies missed in previous inspections.

Like all professional organizations, "Third Party" Inspection Companies in whatever field of expertise, should have at least one yearly gathering, to discuss problems and procedures. This would help the industry.

If I was the Fire Chief or Safety Officer, in a department with multiple apparatus, I would be more concerned about a thorough inspection, rather than the number of apparatuses the inspector could inspect in a day, (for a lesser fee.)

In the mid 1980's, I had a special request from a Fire Chief in a small city in the Northeast. He heard he could purchase a used one-hundred-foot aerial ladder from the New York City salvage yard for a few thousand dollars. There were about fifteen of those aerials in the yard at that time. I was not available to assist the Chief at the scheduled time he was looking to purchase a one hundred foot 'used' aerial ladder. I advised the Chief that I had viewed most of these aerials while working with other fire departments in New Jersey, who were also looking for 'used' one-hundred-foot aerial ladders from New York City salvage yard. I found only one of the fifteen aerial ladders that possibly would cost less to refurbish. All the apparatus I viewed were the light-duty style, one-hundred-foot aerial ladders; and all needed to be returned to the manufacturer for some type of major overhaul. All the apparatus in the yard had excessive mileage on their chassis. Light-duty one-hundred-foot aerial ladders travel countless miles over the road with the ladder bedded on the travel support. The bottom of the extension section rails pounds up and down on the extension rollers. It takes approximately fifty to sixty thousand miles of road travel for the bottom rails to wear and shatter on all the extension sections on these light duty aerial ladders.

Another problem was the turntable bearing. Road travel will cause the forward and rear bearing balls to pound inside the bearing race, thus causing excessive up and down movement of the turntable while operating the aerial ladder. Next, the turntable bearing mounting bolts on most of these aerials were very difficult to access to check for looseness or breakage. Manufacturers that build these light-duty aerial ladders store the ground ladders under the turntable in the center of the chassis. The ground ladders must be removed before the lower side of the turntable mounting bolts can be checked. I have found these bolts loose and broken due to improper torque and from over-the-road travel. They should be inspected annually. Sometimes, it is easier for the inspector to say the bolts were checked, than perform the inspection.

Later that year, I returned to the fire house in the Northeast. The Chief showed me their New York City repainted aerial ladder that had been inspected and a Certificate of Compliance was issued by a another "Third Party" Testing Company. I had already viewed this aerial in the New York City sales yard, so I asked the Chief to what depth had they refurbished the aerial. He explained their mechanic cleaned the unit, and then had the apparatus repainted to the town colors. The Chief stated the fire department only had about fifty thousand dollars invested in this one-hundred-foot aerial ladder, and I must admit, the aerial looked very clean. I asked the Chief if I could take a quick glance at the aerial in the back of the fire house, to check a couple of critical areas on the ladder sections and turntable. We set up the aerial at the rear of the fire house, removed the ground ladders from the storage tray under the turntable bearing, positioned the aerial ladder over the rear of the chassis, extended the sections, and supported the tip. I began to visually check the apparatus. It didn't take long before I was able to show the Chief the cracked and broken areas, on the underside of the ladder extension rails. Then, I crawled under the turntable, through the ground ladder tray opening at the rear of the chassis. From the lower side of the turntable bearing, I handed the Chief three broken turntable bolts, and showed him other missing and loose bolts in the turntable bearing. This aerial ladder had over sixty thousand miles of road travel, per the odometer reading in the cab, and the turntable was very close to failure. The result was the aerial had to be sent back to the original manufacturer for major repairs. The fire department had to find new funding to pay the manufacturer for a new ladder and turntable bearing.

This was one of many examples where 'used' apparatus was tested by a "Third Party" Testing Company, and they did not perform the proper inspection, and simply issued a Certificate of Compliance. The inspection company was sued by the fire department, went to court, and the Inspection Company lost their case.

'Used' apparatus sales companies were everywhere in the mid 1980's, when the heavy-duty manufactured aerials were just hitting the market in sales. Available to fire departments that had no aerial apparatus, was an abundance of older light-duty aerial ladders that were 'used,' or sold as is or, manufacturer's trade-ins for new ladder replacement. The new heavy-duty aerial ladders and aerial ladder platform designs had prices soaring to unbelievable heights. This was the start of a new era of manufacturing heavy-duty aerial devices, and Fire Department Purchasing Agents having to justify the new price increases.

There was a refurbishing company on the West Coast, who purchased a 'used' articulating platform, to make a quick sale to a city fire department just south of the U.S. border. After the delivery of the apparatus, the fire department wanted to put the unit into service, but only after a thorough safety inspection was completed. The company where I was working, prior to when I started PTS, was called by the Fire Chief to come, and inspect this aerial platform. I was sent down to the Border to perform an inspection on this unit. There was an interpreter on location to make it easier for them to understand the inspection. The unit was brought to a large vacant lot. The power takeoff was engaged, and the outriggers were set. I was ready to operate the aerial platform. I went to the control station at the turntable to begin operating the platform. Just then, the interpreter started yelling for me to stop operating the controls. He explained to me there was an auxiliary hydraulic pump in the front of the body, and they were told it would have to be started in conjunction with the regular hydraulic pump before the aerial platform could be operated. This procedure was to insure they would have plenty of hydraulic power for the aerial platform. That was the operating procedure given to them by the delivering representative for this unit. Working through the interpreter, I finally convinced the fire department mechanic that it was O.K. to operate the aerial with the main engine pump only; and only if the truck engine stops, they should use the optional auxiliary pump in an emergency to operate the controls to store the aerial for road travel. I asked the interpreter to tell the mechanic to check the auxiliary pump operation frequently. Now, I could continue checking out the operation, with the truck pump only. When I started to lift the upper boom, I noticed that the platform was not leveling properly. I quickly let off the raise control and lowered the booms back to the travel support and inspected the platform hinge pins. The pins and bushings had excessive salt rust pits and were frozen together. The Chief said since delivery, they have parked the aerial platform on a lot near the ocean for the last three months, as there was no room in the firehouse because of broken down equipment strewn all over the floors. The platform was eventually repaired, and the unit was put back in service later. The aerial platform was then operated with only one pump at a time. This is another example of how one of those 'quick sale' companies did not know the proper operation for the equipment they were selling.

Similar Platform - South of the Border
Picture courtesy of Reliance Fire Museum, Estes Park, Co.

There are many fire departments, not only in the United States but out of the country, that do not have funding for equipment. Many must wait months to pay for minor breakdowns of their equipment as was this department south of the border.

An annual inspection was scheduled to be performed on one of the first 1970's Grove Aerial Ladder still in operation. The ladder had chain and cable extension. This ladder had many years of workout in training and emergency situations. PTS was called to do an annual inspection and certification for the first time. While inspecting the ladder guides at the tip of the base section, I noticed they were worn to the extent that full horizontal extension could result in the sections coming apart. This problem was discussed with the Fire Chief, and the chief followed up with the manufacturer. Now, this is where the problem began. The chief was convinced there was a problem with the base ladder section guides. The manufacturer decided since they were close to where the fire department was located, they would simply send a service representative to check out the aerial ladder situation. The service representative made the decision the aerial ladder was not in any danger, and for the chief to let it remain in service. Since earlier I had showed the chief the problem, he was not convinced that the service representative had given him the correct advice, so he called me back. I called the design engineer at the factory. I had dealt with this engineer previously, and personally knew him. I explained what had been found on the ladder, and he replied that he would personally check the aerial ladder out himself. The engineer went to the fire department; and after checking out the situation, he agreed with our original determination, so the ladder was confirmed defective, and quickly sent to the factory for repair.

In the Southeast, a fire department was having an annual inspection on one of their light-duty one-hundred-foot aerial ladders. During the inspection, only minor items were noted. Later, the chief called and said the "Third Party" Inspector had missed a cracked spring leaf in the rear of the chassis springs. The chief said he contacted the manufacturer, and they sent a service representative to check out the problem. This aerial had hydraulic outriggers that could lift the chassis off the ground, mounted in front of the cracked spring leaf. The service representative advised the chief the aerial was unsafe for use because the chassis could tip over if the aerial was rotated to that side, and the chief was to remove the aerial ladder from service. This is another case where an inexperienced manufacturer representative gave bad advice. It was finally proven that even with a cracked spring leaf, the aerial was still structurally safe to use.

The point is, mistakes are made so do not assume that when a decision is made, the one making the decision is always right. You should feel very confident when operating any apparatus. Sometimes, more than one decision is necessary to satisfy a situation.

Today, it's much harder to sell the older aerials, due to the amount of new style much improved models. Insurance ISO, ratings dictate and make it difficult for the smaller departments that have no aerial apparatus to qualify. We may have a lot of antique apparatus around.

LIGHT-DUTY AERIALS/PLATFORMS

FAILURE AND INJURY

Throughout the years of aerial manufacturing, there have been many wooden and metal aerial ladders that have failed, while using them for some type of rescue or special training operation. Operators, who have not been properly trained in setup, or how to operate the apparatus, were responsible for many of the aerial failures. Short-jacking, or forgetting to set the outriggers properly, not using outrigger ground pad, when necessary, has resulted in many failures. When the unit was short-jacked, and the aerial was rotated over that side, failure and injury has resulted. There were times, when the ground surface was unstable, and failure of the apparatus would result.

Not properly setting ladder extension locks, on medium-duty ladder sections can be serious. Sometimes at full extension and full elevated ladder sections have slipped inward, resulting in injuries for ladder pipe operators and personnel at the ladder tip. The ladder sections can retract for some reason or other, catching the feet between the rungs, breaking both feet. Most fly ladders have fold-out footsteps at the tip, but it was still easy to have toes caught between the rungs.

Resting the fly ladder sections on window ledges or supporting the ladder tip during a rescue, was a false sense of security, but was always recommended as SOP. It makes the ladder sections have better support, than in an open cantilever position. In the early years, operators of apparatus had no knowledge that even though the lift cylinder locks were applied, the lift cylinders could have some minor internal leakage, and the ladder sections would lower into the window frame, causing reverse overloaded stress on the sections. Failure of the ladder sections could and did result, during rescue operations. Overloading the sections during rescue was a common event. The ladder sections could have been under undue stress before rescue, and the operator would never perceive failure would happen. These light-duty ladders were not designed to be fully extended in the horizontal position unsupported, and have an overload applied at the ladder tip.

Picture courtesy of Fire Engineering

Reaching out over water to rescue a distressed person, or where a long reach was required to lift a weight, ended up in many aerial failures. When only the base ladder is used for lifting, there was less danger of failure, as demonstrated by the Canadian Manufacturer Pierreville Fire Trucks, Inc., in lifting a Volkswagen Beetle from the tip of the base ladder.

Overhead power lines play a big role in many ladder failures and injuries. There have been ladders and platform failures caused from simple field operations and routine tests.

Platforms are supposed to be manufactured with a one and one-half to one safety factor, for platform capacity. I could never understand when doing a load structural test on an aerial platform, at one and one-half to one, why the "test load," as recommended in the manual, was to be placed inside the platform prior to operation. On many aerial ladder platforms, a most unstable position for the platform is at forty-five degrees angle front and rear of the chassis. If a platform had the "test load" placed inside the platform and was rotated to forty-five degrees, at full extension reach; and the chassis became un-stable, the ladder platform could fail. Not having proper outriggers reach for the aerials they were design for, has promoted many unstable conditions.

Here are some incidents fire departments in various parts of the country had and most were considered operator error. Not all failures, or injuries, were the operator's fault. There have been incidents, where "SOP," Standard Operating Procedure, by the fire de-partment, has related to injury and death.

Pennsylvania Volunteer Department:

It was very late and dark out in the countryside; there was a large working house fire. The volunteer fire department had a one-hundred-foot light duty aerial ladder that was to be elevated and extended for water pipe operation. When the firemen arrived at the fire, the apparatus driver set the unit up as close as he possibly could to the fire. the ladder operator then raised the aerial to maximum elevation and extension, when he rotated the ladder toward the fire, the tip section touched overhead electrical wires. The electrical wires started arcing hi-voltage into the aerial tip to the tires and ground. At this time the ladder operator was safe and not grounded. This fire department had an "SOP," and if the ladder gets into this situation, the ladder must be manually hand cranked rotated away from the wires. The hand crank was not at the upper control station but stored in one of the lower body compartments. Following the departments SOP, the nervous ladder operator jumped from the turntable of the aerial to the ground to get the crank out of the compartment. When he touched the compartment door, the voltage was electrocuting him. An EMT technician was standing close by, he threw a body block into the operator, and knocked him away from the compartment door. The EMT technician saved the ladder operator/firefighter's life. Another firefighter with aerial experience, jumped onto the chassis, and rotated the aerial out and away from the wires.

Arkansas Paid Department:

This paid fire departments first line apparatus was a fifty-foot articulating aerial platform. Every morning the aerial was to be driven outside the fire house bay as per the fire department's SOPs, setup and operated through its functions. At the end of the driveway apron, there were many overhead electrical wires running parallel to the street. A semi-experienced fireman/operator, got up early one morning, and thought he would do this chore for the day. He drove the platform to the end of the apron, set the outriggers in place, climbed into the platform, and proceeded to raise the booms upward. The operator had set up the aerial at the end of the apron, and too close to the street. While elevating the platform, the platform railing became very close to the electric wires. The operator was in the platform, and reached out with one hand, to move the wires away from the platform railing while his other hand on the platform controls. The operator tried to continue elevating the platform. As soon as his hand touched the wires, an electrical surge went completely through his body, blew out the tires of the apparatus, and made quarter-size blow holes through the outrigger steel foot plates, where they contacted the ground. Another firefighter was observing this predicament, acted swiftly, and jumped on the rear step of the chassis. The firefighter was able to get to the platform lower controls to override the upper controls and lower the platform away from the arcing wires. This firefighter put his own life in jeopardy, but he saved this firefighter's life.

Delaware Volunteer Department:

It was a beautiful sunny Sunday afternoon, skies were clear, with temperatures around seventy degrees. The members of this volunteer fire department were operating their light duty one-hundred-foot aerial ladder in an exercise and training program at their facility. The fire chief's son 200 lbs. plus was in charge; and after the training exercise finished, he wanted to show the other volunteers present, how he could hold onto the tip of the fly section at a horizontal, fully extended position, and be rotated around the chassis. Another volunteer firefighter was to operate the ladder controls while the chief's son hung from the tip of the fly section. The operator rotated the aerial quickly around the chassis; when he stopped rotating, the chief's son's momentum continued to whip the fly section to the side causing it to twist and collapse with force. The chief's son fell to the ground, and he broke his back. This was a very foolish exercise and should never have been permitted.

Connecticut Volunteer Department:

A call to a house fire at night, required the light-duty one-hundred-foot aerial ladder to be elevated and extended, close to electrical wires for ladder-pipe operation. There was ample clearance for aerial operation between the wires and the house fire. The water hose was placed in the center of the extended ladder sections, and the water hose and water pipe were strapped to the tip of the fly ladder section. There was a firefighter at the ladder tip to control the nozzle operations. As the water pressure was applied to the water pipe, the tie-down strap broke, and the nozzle and hose lifted away from the tip of the fly section ladder rung. The firefighter was not wearing a safety belt, he was straddling the hose, and was lifted off the elevated ladder. The firefighter was launched backwards and bounced off the electrical wires. He held onto the water hose, as he fell to the ground. The "reason" the ladder pipe came off the rung was the locking device was defective, and the tie down strap broke. This strap was always hanging at the rear of the chassis where it was exposed to excessive sun ultraviolet rays, which weakened the material in the strap. The firefighter should have been securely fastened to the tip of the ladder section before the water flowed. The firefighter lived through this experience, but he was severely injured

Arizona Volunteer Fire Department:

There was a similar incident to the Connecticut fire department. A volunteer fireman was standing on the footpads, at the tip of a fully extended and fully elevated, light-duty sixty-foot aerial ladder with electrical top controls. This aerial ladder operation had extremely quick movements. All functions, raise-lower, extend-retract, rotate, were less than twenty seconds to achieve. This firefighter had on his safety belt, and it was attached to the top railing at the aerial ladder tip. This ladder also had dual controls, one set at the base control console, and electric pushbutton controls at the tip of the fly section. As he was going through the aerial motions using the electric full on-off controls at the ladder tip, he was raising the aerial upward to the elevation of seventy degrees, he released the elevation control button. The ladder sections whipped forward and backwards after releasing the control button and ripped the fireman's safety belt apart. The firefighter was launched off the ladder, and he fell sixty feet to the ground. This operator's safety belt was always anchored at the rear of the apparatus for quick access. Like the strap on the ladder pipe, this belt had excessive sun ultraviolet ray damage, from hanging on the rear of the apparatus, so again the material weakened and failed. It was unbelievable, and a miracle, that this firefighter survived. I spoke with him, and he told me that all he could think about, while falling to the ground, was to try to land flat on his rear end and keep his head up.

The short summary above gave examples of 'real life' happenings, where various accidents and failures innocently occurred mostly with the lite-duty and medium-duty aerial ladders.

SAFETY BELTS:

As a matter of note, there is a procedure for inspecting and testing safety belts, there is a procedure developed and tested by the manufacturer. It is almost as costly to send the belts back to a manufacturer to get inspected and stress tested, as to buy a new safety belt. none of the safety belts stated above have ever been properly tested. PTS developed a procedure for 'stress testing' safety belts, as per manufacturer recommendations. These inspections were performed on only a few safety belts at fire departments that also had additional apparatus. Safety belt inspections became a catch twenty-two, but there is a need for this to be done. The inspection cost, versus replacing the belt, was nearly the same, especially when some departments did not have other inspections to perform at the same time. Therefore, safety belt 'stress' tests faded away to only a visual inspection, and with recommendations to the fire department to be aware and inspect or replace more often. Has your department ever been reminded to keep a vigil on all safety belts, and check frequently for any damage, by a "Third Party" Inspection Company? Ultraviolet damage is real, and the material will fatigue with sunlight rays. Many belts are attached to the rear of the apparatus chassis and hang there, year in and year out. Safety belts need to be inspected and kept out of sun rays as much as possible.

AERIAL LADDER/PLATFORM:

ROAD DAMAGE

There are many ways aerial ladders, and aerial platforms parts **wear out** or **fatigue**, to the point of being unsafe to use. The biggest culprit to deal with is "over-the-road" travel. Roads do have dips, ruts, and rough areas that will shock the apparatus while traveling to destinations. Aerials are usually clamped or tied down one way or another, to some type of travel support. Some travel supports have shock absorbing material installed, to cushion the up and down road shock pressure, and many have locks, that help grab the ladder or boom sections, to keep them from bouncing around. Still damage results to the ladder and boom sections, regardless of what type of support. Unforeseen damage to the aerials can occur from the operator not storing the aerial properly.

Aerial ladders were first designed with rollers to support rails and assist with the extension of the sections. This was a good idea at that time, but they had no idea that after long time usage, damage was imminent. There were various types of support rollers used at first. One manufacturer made the rollers from wood and phenolic materials. Other manufacturers used steel or aluminum for rollers. The apex of these rollers is very small, thus when the apparatus traveled over the road, there was shock transferring from the weight of the ladder sections, bearing down on the rollers. After many miles of travel, from here to there, the material on the underside of the ladder section rails would fatigue, crack, and shatter to the point that some rollers apex could fit inside the shattered rails. The phenolic rollers would put flat wear indentions on the underside of the rails and was softer material than the steel or aluminum rollers, so fewer cracks to the rails were found on that type of ladder, and the rollers usually wore out first. When the rollers would get gummed up from contaminated grease and dirt, The rollers would not spin, so the apex would flatten out quicker, as the rollers malfunctioned. Manufacturers would keep replacing the worn and damaged rollers. It was many years be before the new heavy-duty aerial ladders were introduced with slide blocks to replace rollers. Aerial ladders found to have this severe roller and rail problem, usually had over fifty thousand miles of road travel. Damage was caused mainly from the steel rollers supporting steel ladder sections and the over the road travel causing shock to the sections at the roller supports. Before "Third Party" Inspections evolved, some fire departments would send these damaged ladder sections to local welding shops, who welded gloves to cover the damage areas as a quick fix. These ladder rails are made of very light gauge high tensile material; and most of the time, the welding shops did more damage to the rails, than repair them. In the long run, the ladder sections had to be replaced by the ladder manufacturer. To eliminate the roller problems, slide blocks were finally introduced by one aerial ladder manufacturer. It was not until years later when all light-duty aerial ladder manufacturers were replacing rollers with slide blocks, for better support of the extension rails. New style heavy-duty aerial ladders have slide block supports for ladder extension and rail support. This modification increased safety and longevity for all aerial ladders.

The weight of the aerial ladders and platforms created another problem while the apparatus was traveling over the road. The problem was the travel support shock, was carried back to the turntable bearing races. The constant road shock would wear the front and rear bearing races, from just applying hydraulic down pressure while storing the base ladder or lowering the boom onto the travel support. These bearing races are filled with ball bearings, and are assembled with little clearance, for smooth movement and stability. The road shock would make the ball bearing apex wear into the race, causing excessive up and downplay when the aerials were operated. It only takes a few thousandths of an inch, of up and down wear, inside the bearing race, to make an unsafe aerial. I have seen articulating aerial platforms, when the lower boom section is raised above vertical, shifting its weight from front to rear, causing the platform or basket to free fall up and down just from the minimal worn turntable bearing. (This turntable bearing internal wear is checked with dial indicators). When this happens, the upper boom platform would free fall as much as five to fifteen feet, according to the amount of turntable bearing race wear. This play would cause undue stress on the mounting bolts, and they would loosen, stretch, and sometimes break. Articulating platforms were always supported by some type of travel support mounted to the chassis frame. These booms are very heavy; and even though they are supported for travel, they still bounce and move on the support, while going over the road, this would cause metal fatigue and cracks, to the underside of the boom sections where they rest on the support, and to the travel support frame. Many of the larger articulating boom mainframes would crack on the underside of the boom that rests on the travel support. Some cracks were from six to twelve inches in length. Most of the travel supports were found with some type of damage, that had long over the road travel. Over the road travel is one of the most destructive aspects to aerial equipment failure. Shock through all the components can cause welds to crack, bolts to shear or loosen, and material to wear and fatigue. Not only do the aerials take a beating, but the chassis and body components wear out quicker, from the stress and shock. Travel supports for apparatus, and ground ladder supports, still to this day, cause excessive wear to the equipment, from over the road shock and vibration.

MAINTENANCE:

Larger fire departments usually have their own maintenance service centers to maintain fleet apparatus on a regular basis. The complexity of the equipment requires the department mechanics to spend more time inspecting, and servicing. When they are short-staffed, they are unable to complete routine service on all apparatus. There just isn't enough time in the day to do everything. Sometimes this leads to a scheduling and servicing problem. Not having mechanics on location, the smaller fire departments must send their apparatus to a service center or to the manufacture for service. This keeps apparatus out of service a long time. Some departments will have a service man come to the firehouse to do light duty service on the equipment. When this happens, it's difficult to get a one hundred percent service, and areas on the apparatus, that should be checked and maintained, are sometimes skipped.

Moonlighting:

There are fire department mechanics that moonlight and help smaller fire departments that have budget problems and cannot afford to send the unit to the manufacturer for service. Many times, this takes a unit out of service for a long period of time, while waiting for parts to be ordered, or whenever the mechanic can get back to the job. There are pros and cons about repairing equipment in this manner, in the fire house bay. One thing to be aware of is, these fire equipment mechanics are usually doing the extra work to help their income, and probably do not have any insurance coverage to cover their work. This doesn't mean that they are not fully qualified to help. Care should be taken to be sure the mechanic is aware of the fire department's policy, in case of an emergency.

In House Exhaust Systems:

In the early years of converting fire houses from horse drawn apparatus to accept steam engines, later to convert the fire house to accept the latest fuel powered engines and aerials has been an ongoing event when new equipment is purchased. Many Fire Departments have an in- house mechanic-fire fighter, that does monthly service on the apparatus. Some larger Fire Departments have Maintenance Service Centers with full time mechanics to perform service on equipment. There was always a problem of servicing the equipment inside the fire house or at the maintenance centers, especially when it was cold, or weather did not permit working outside. The problem was the possibility of poisonous gasses emitting from the exhaust system when the equipment engine is started. It wasn't until late in the 1900s when an innovation to eliminate engine exhaust in the fire houses was developed. A flexible hose attached to the tail pipe end would grasp the tail pipe and only release after the apparatus is driven out of the fire house and passed the door entrance. The hose was connected to an exhaust fan at the top of the fire house roof, and this eliminated the poison gas that would harm the firemen, or the mechanics health when the apparatus is started and idling inside the fire house. There are many small volunteer fire departments with limited equipment that operate with moonlight mechanics, in the old fashion way.

Brakes, Retarders, Limiting Valves:

In the early 1990's, the National Transportation Board made recommendations, that the nation's fire departments act, to improve the safety for their vehicles, getting to and from emergencies. The recommendation included enhancement of vehicle maintenance programs, mandatory and enforced use of occupant restraints, and procedural changes in use of some mechanical equipment. This recommendation was based on an investigation of eight traffic accidents involving fire department vehicles in 1989 and 1990. From and during the 1980's, there were one hundred seventy-nine firefighter deaths, where fifteen percent occurred because of accidents involving fire apparatus in route to emergencies. The board expressed that most of these accidents were from lack of proper maintenance and inspection programs for these vehicles. The Safety Board expressed extreme consideration to vehicle brakes. In the Northeast, there was an accident where two firefighters lost their lives, because of the condition of the brake canister slack adjusters on the apparatus were not properly adjusted. The brakes had about fifty-eight percent of their original braking capacity. It was not stated if the brake pads were severely worn, or brake slack adjusters out of adjustment. Following that accident, the Safety Board conducted a voluntary inspection program for fire department service vehicles. It was very surprising, that there were thirty-one states that did not require fire service vehicle inspections. During this inspection program, there were five hundred fifty-nine pieces of fire apparatus checked from sixty-four cities throughout various states, and results found that over thirty-five percent of the vehicles checked, should have been placed out of service. The Safety Board also addressed engine retarders, and brake limiting valves used on fire apparatus. The National Safety Board found that engine retarders on fire departments apparatus, were routinely used, even though manufacturers have cautioned against their use especially on wet and slippery roads. The Safety Board urged fire department associations, to inform fire departments personnel and firefighters, of the dangers of misusing retarders. Manual brake limiting valves have a purpose to reduce the braking capacities of the front axles, and to reduce skidding of the front wheels. The investigation showed that manual limiting valves only reduced braking power of the vehicle, while not reducing the likelihood of skidding. The Safety Board's investigation of the Northeast fire department's accident stated that the limiting valve would have reduced the already poor breaking power of fifty-eight percent, to thirty-six percent of the total breaking power of that vehicle. The result of the investigation was a recommendation by the NTSB, that manual brake limiting valves should never be used, under any circumstances, and mandate a proper inspection and service of the braking system on all vehicles be enforced.

Brakes, Drums, & Slack Adjusters:

Fire departments with tractor-drawn aerials have more trouble with the braking system, than aerials mounted on single axle vehicles. The drums at the fifth wheel duals seem to take most of the breaking torque and are more susceptible for damage. During a routine inspection, drums have been found in service, with stress cracks wide enough to insert a knife blade. There were drums in service, that had cracked apart and were about to explode. On tractor- drawn aerials, the brake slack adjusters were found to be one of the main problems, and usually were out of adjustment. When brakes were applied, the fifth wheel drums would take most of the force to stop the vehicle. This caused the drums to wear out more quickly. The front brakes should be the last to apply, keeping the force off the front. When all brake slack adjusters are properly adjusted, the tiller axle is first to start applying; then secondly, the fifth wheel starts applying, and the front brakes apply last. Not only do the brakes last longer, but also do the drums. The brake slack adjusters will not have a great difference in stroke but should be in a sequence to apply and reduce damage. At each regularly scheduled maintenance inspection, the drums, pads, and slack adjusters should be checked for damage, wear, proper adjustment, and loose mountings.

Steering Arms:

In the mid 1980's, after a few accidents had happened, an alert was issued to inspect and replace chassis steering arms with designated serial numbers on large trucks from a particular manufacturer. A problem was caused from over-the-road shock to the front steering arm on the apparatus. The steering arm would crack and sometimes break in a certain area, due to the way the arm was manufactured. When this happens, the arm breaks in half and all steering to the vehicle is lost. There were many units in service from this manufactured with this steering arm, and all had to be called in for a special inspection and removal for testing and possible replacement. The defective arms were checked, using a field portable magnaflux unit. Almost every steering arm inspected with that serial number, had some type of crack indication. One fire department had over 30 of these units with damaged steering arms. An 'extra heavy-duty' steering arm was reinstalled on all units with this serial number if a crack was found or not. This red flag was flown in time, which reduced future problems and possible injuries. This is just another area for maintenance personnel to visually check when servicing the apparatus.

Vehicle Promotions:

Oil companies like to promote their products, especially in larger cities with many fleets of various type vehicles and apparatus. This was a great idea that not only cuts expenses for cities, but also benefits the employees. The oil company would give all the mechanics and fire departments, free cases of oil, along with oil filters, and supply the city with enough products to service all vehicles and apparatus. This promotion was presented to one of the larger cities in the south, to try for a period. Advertising and TV ads of this product line with this city were abundant. Not long after the changeover to this product, many of this city's vehicle engines began to break down and malfunction. After an investigation, it was concluded that the oil used was the main problem. This city discontinued their contract with the oil company. They canceled the TV commercials and returned all the reserve oil to the manufacturer. Shortly after removing the oil from the vehicles, the engine problems for that city ceased. Quite a mystery, one wouldn't think could happen. PTS had inspected apparatus at this city for approximately twenty continuous years. Through these years of training personnel and working with their mechanics and inspectors, there was never an apparatus failure due to one of our inspections, other than routine breakdowns or unrelated accidents.

Lubrication Devices:

There is a fire department in the eastern part of the U.S. that helped a manufacturer develop an automatic lubrication device for fire apparatus chassis. This system provides lubrication to at least twenty-two points on the chassis. This system proved very efficient and cut down maintenance on equipment by a large percentage. The only service to this system was to make sure the lubricant container was full of lubricant, and to do visual inspections of the lines and fittings for wear or damage. This was one tool found to be worth every cent invested. Manufacturers of apparatus have installed this system on many of their newer units. This lubrication kit is still available and used by fire departments everywhere. The kit can be installed at the fire department maintenance shop by the in-house mechanic, fire apparatus repair facilities, or at the manufacturers

Counterfeit Bolts:

In the late 1970's and throughout the 1980's, there was a huge problem that took many years to make aware of. The problem was an influx of counterfeit or substandard fastener's "Nuts and Bolts" that were distributed throughout the U.S. and shipped in from foreign countries. Distributors were selling this imported merchandise with our SAE national standard markings on the heads of the fasteners, along with markings from wherever the fasteners were manufactured, to every city maintenance shop and manufacturing business in the United States. There were many failures of these fasteners in every industry. It wasn't until the early 1990's, when these fasteners were finally considered "counterfeit," by the Defense Contract Management District South, and the National Highway Traffic Safety Administration. There was also a Safety Bulletin released in 1992, by the National Institute of Emergency Vehicle Safety, advising fire departments and other emergency response agencies, to check vehicles for counterfeit and substandard bolts, manufactured prior to the 1990's. The EVS board investigators identified 'suspected' counterfeit fasteners on one fire department's apparatus, as well as in the maintenance shops bolt bin. Manufacturers of fire apparatus began to replace the counterfeit, or 'suspected' counterfeit bolts, found on apparatus already in service and at their manufacturing facilities.

These counterfeit fasteners were first exposed by The National Highway Traffic Safety Administration, with reference to the crane and derrick field, and through the trucking industry. Shortly thereafter, the (NHTSA), found that fire service vehicles, also had these 'suspected' counterfeit fasteners. Not only did the manufacturers have an influx of counterfeit fasteners, but city and county municipal repair garages were stocked to the brim with these fasteners. As the counterfeit bolts were removed from manufacturers and municipality bolt bins and replaced with properly marked fasteners made in the U.S., there were many units in the field that had to be checked. We found counterfeit bolts mounting components to the chassis, steering devices, driveline, mainframe mounts, and just about everywhere a fastener was needed. On aerial apparatus, counterfeit fasteners were also found in the turntable bearings, along with the frame and travel support mounts. Manufacturers of apparatus issued special instructions for inspection and replacement of counterfeit bolts found on their products, and how to replace and properly torque the new U.S. manufactured SAE bolts. It took a long time working with the manufacturers to get most of the counterfeit fasteners out of the system and off the field units. There were aerial failures during the period from the 1970's through the 1980's, which were not specifically blamed on the counterfeit bolts, because of limited knowledge about this vast problem. In 1987, after a Congressional investigation was launched, the Fastener Quality Act, Public Law 101-592 was passed. Even though the problem was addressed on a national basis, the Congressional investigation seemed to bypass the emergency response community at that time.

Working in California as an accredited inspector in the mid 1970's, I was inspecting cranes and derricks for power companies, along with other municipalities. There was an instance where a two-hundred-foot-high horizontal crane was being assembled for the construction of a second power plant substation. The crane was in its final stage of assembly. The counterweight section had just been installed to the top of the mast, seven workers were placed on that section to work on the winch and wire electrical components. The workers were preparing to attach the jib section to the top of the two-hundred-foot vertical mast. A mobile lattice-type crane was being used to support the vertical mast at the tip. The jib section was lying on the ground ready to be lifted to the top of the mast for attachment. The workers on the counterweight section discon-nected the cable support at the top of the mast, so the mobile crane could pick up the jib section and lift it into position. As soon as the workers disconnected the tip support cable, mounting bolts on the counterweight and mast sections started to shear. The nuts stripped off the mounting bolts, causing the mast to fold and topple to the ground. All seven workers died. This accident was in the mid 1970's, and it was never disclosed whether the bolts used for the construction were the possibly cause of the failure, or if the seven workers were just careless. This was a dual tragedy, with both inferior products being used, and workers creating this situation.

Influx of counterfeit fasteners was at its peak, and there were many distributors of these counterfeit fasteners in the United States. At that time, no one thought these counter-feit bolts ever existed or what counterfeit fasteners were. Most of the counterfeit bolts had "KS" marked on the fastener head. These fasteners were manufactured in Kosaka Kogyo, Japan, and other sub-substandard fasteners came from Taiwan and Canada. Inexpensive counterfeit bolts are presently being sold over-the- counter in hardware, lumber supply, and discount stores everywhere. If the mounting of a product requires high tensile strength bolts, one should research and purchase from a replicable dealer or manufacturer, and 'always' get certification of the bolt strength, and carefully torque to required values only.

TORQUE TURNTABLE, CHASSIS BOLTS:

There is fire apparatus where extensions to a torque wrench must be used to check mainframe and turntable mounting bolts. Many of these bolts are very difficult to check, because of limited access. During installation at the manufacturer, additional body and associated equipment was added, making it more difficult and sometimes impossible to check for loose or broken bolts. When long extensions are used with torque wrenches, accurate torque values are not always achieved due to the twisting and flexing from the extensions. All that can be determined, when long extensions of more than six to eight inches are used, is if there are any loose or broken bolts. Extreme caution when applying torque with long extensions is advised. There are smaller turntable bearings on fire aerials, that have different sized socket head mounting bolts top and bottom of the bearing. That's when trouble begins. There are two different torque values for these socket head fasteners; and if the torque value on the torque wrench is not changed and bolts are checked with the higher value, the smaller bolts can break or threads strip in the bearing. There are turntable bearings that are mounted with grade 5 bolts, and extreme care must be taken to not over torque these bolts. There are mounting bolts that require a horizontal extension wrench to be added to the torque wrench, to reach in narrow areas. This extension adds torque to the torque wrench setting and must be recalibrated to offset the extension. Always check a manufacturer's bolt chart for proper values before attempting to torque bolts, and make sure the torque wrench used is properly calibrated. Then finally, there are mainframe and turntable bolts that require a torque multiplier, because the torque wrench will not adjust to the higher torque values required. Some of these bolts' torque values are over five-hundred-foot lbs. of torque. It would be easy to break or strip a mounting bolt with the multiplier, so proceed with extreme caution and do not exceed torque values.

Component Frame Chafing:

It is amazing how vibration and over-the-road shock will chafe hydraulic and electrical lines that are simply lying across or bending around the frame or supported by other chassis members. One item is the large hydraulic line that goes from the hydraulic reservoir to the pump. This hose seems to chafe and break steel ply when not properly supported. It is much easier to inspect all these items while the apparatus is up on a maintenance lift. Visually observing electrical wiring and hoses at every scheduled service, to make sure they are not in a condition to rub or cut on frame edges, will help reduce emergency breakdowns in the future. When first assembling apparatus components, manufacturers will install some type of padding or frame cushion component to prevent chafing of hoses and electric wires. These components sometimes fall from the frame, causing them to rub and chafe.

Drivelines:

Universal joint driveline bolts and bearings should be checked for wear and loose or missing bolts at every service. We have found many universal mounting bolts missing and bearings worn, due to lack of lubrication. Most of these deficiencies were found on units at smaller fire departments that have no scheduled maintenance program, or personnel to perform the task.

Stabilizers:

Outriggers, especially telescopic H-type, need to be checked for hydraulic hose damage, and telescopic housings for structural damage. The hoses travel back and forth inside the horizontal extension sections, and rub on the inside of the housing, at times jamming and breaking. The vertical telescopic housings with holes for manual locking pins, which prevent retraction of the housings, will get structural damage when the cylinder is retracted, and the pin is not removed. Many of these housings were found unrepaired or replaced, and the locking holes are so elongated, that the safety pins would not fit into the elongated holes. The locking pins are a safety precaution in case the extending cylinders have internal leakage or some type of failure. After setting up the h-type outrigger system, while operating the aerial around the chassis, you may hear a popping noise from one of the outriggers. Every time that noise is heard, one of the outrigger cylinder shafts is retracting from internal seal or locking valve hydraulic fluid leakage. This cylinder shaft retraction usually is very minimal and measured in thousandths of an inch. This drifting can be determined with dial indicators attached to the extending housings. This doesn't sound like a lot to worry about; but over a period, the housing can retract enough to lock onto the safety pin. When this occurs, the hydraulic outrigger control must be actuated to release the pin from the outer housing. If the telescopic outrigger housing retracts quickly back onto the safety lock pin, the cylinder or holding valve may need repair or replacement. Never depend on safety lock pins to support the apparatus but be sure they are all set in place prior to operating the aerial.

A-frame outriggers will get overextended sometimes and will not retract into the housing for road travel. When the this happens and the booms are stored on the travel support, they rest on top of a hydraulic transfer switch, so the lower outrigger controls are functional, but the outriggers still won't retract and are not in road travel storage. To get the outriggers to store and retract requires the lower boom raised off the travel support, rotate the lower boom to one of the chassis sides, just enough to relieve pressure on the opposite side outrigger. Manually depress the boom travel support switch lever to actuate the outrigger controls and slightly retract the opposite side outrigger to break the locking force. Reset the boom to the travel support, and the outriggers should retract and store for travel.

Chains & Cables:

Cables and chains used for extension of ladder sections or platform leveling need to be checked for proper adjustment, stretching, worn flat areas, and broken strands or links. Cables can get out of alignment by simply adjusting or tightening the mounting nuts. When the cables are out of adjustment, extension pressure is on one side of the ladder section and will wear those cables and pulleys out much sooner. When cables rest on pulleys for a period, flat areas can form from metal fatigue and cable strands weaken and begin to break. When three strands of cable in one lay on any cable, are fatigued or broken, the cable needs to be replaced. Maintaining cables sprockets and chains, cleaning and lightly lubricating, and keeping them properly in adjustment, is very important for longevity. A soft cloth wrapped around a cable, and moved along the cable lengths, will find broken strands much easier, than simply viewing or rubbing your hands over the cables. Numerous extension cables have been replaced on aerial ladder sections. Some of the cables were installed and attached with cable clamps. Many of the cable clamps were found to be installed incorrectly, and some cables were damaged and had to be replaced. There is a proper way to install cable clamps. A standard cable clamp has a U-bolt and a clamp body. When the cable is installed and ready to attach, there is a pigtail that attaches to the main running lay of the cable with a clamp. Many times, the clamp body is installed to hold the pigtail to the cable, and the U-bolt will then crush the main cable lay, resulting in cable failure. Always have the clamp body installed on the main cable run, and the U-bolt installed around the pigtail. By standard, it usually takes more than one cable clamp to fasten the cable ends together; usually three clamps are used and equally spaced.

Chains may skip a sprocket tooth, causing twisting to the extending sections. Connecting links and pins can lose their keeper, and sometimes the sprockets will shear a key. Some chain adjusting brackets have broken away from their mounts, and chains get out of adjustment. Chains, as well as cables, need frequent cleaning, and proper lubrication.

Bearings:

There are different types of bearings on aerial apparatus, which require service and lubrication, to keep them functioning properly. Aerials that rotate three hundred and sixty degrees continually have some type of rotation bearing. These bearings have split races, one mounted to the mainframe, and the other to the turret assembly. Inside these races are ball bearings that assist the races to rotate without any resistance. There should be an area that permits lubrication to the internal race and bearing balls, and they should be serviced on a regular basis.

Some elevation or lift cylinders on aerial ladders and platforms have swivel bearings at the mounting ends of the shafts and barrels. These bearings allow the cylinders to function with little resistance. Most have grease fittings for lubrication of the bearing races. Sometimes the lubricant will not penetrate the bearings, the bearings will lock up and damage will result. On the aerials that have these type bearings installed, a good maintenance program to service these bearings is recommended.

AERIAL TRIVIA:

Since the beginning of aerial manufacturer in the 1800's to present day many quirks and problems have shown up in various manufacturer's products. There has been operator error, manufacturer design failure, and maintenance problems that were overlooked, that have caused failure to equipment and sometimes injury to personnel.

Many questions have been asked throughout the years, by mechanics and department personnel, who were puzzled about certain extraordinary operations or functions of their apparatus. Some problems took longer to solve than others, but often departments were operating with inaccurate procedures and lack of proper training.

Here are a few incidents that baffled fire department personnel over the years. They relate to maintenance, poor instructions, or idiosyncrasy problems.

Test your knowledge on these questions about older aerials; many of you probably weren't born yet, but some of the problems still exist. See if you can answer a few of these aerial apparatus questions.

"Third Party" Independent Testing Companies, a manufacturer's quality control representative, or a mechanic before PTS was called, have already done their inspection or maintenance process. Here are a few perplexing situations we encountered.

One question frequently asked by fire department personal, that are confused as to Third Party Inspection Companies, (Is this a UL test?)

The answer is **NO. UL did not invent this testing procedure.**

TRIVIA QUESTIONS: 1. Maxim 1970's - 100 Ft. Aerial Ladder:

Southern California:
A one-hundred-foot Maxium rear mounted aerial ladder, was driven to a regular non-emergency call, when it approached a school bus stopped at a traffic light. The apparatus driver applied the brakes, stopping behind the bus. After stopping, the tip of the aerial ladder extended over three feet, and nearly touched the back window of the

school bus. Then, as if nothing had happened, the tip quickly retracted back to the road travel position. This quirk had never happened to this aerial before. The fire department mechanic immediately called the manufacturer's distributer. A service representative came immediately to the location. The mechanic and the representative test drove the apparatus several times. They tried to determine why the aerial ladder tip extends and then retracts back when they come to a stop. The ladder sections would extend different lengths, according to the speed and how hard the brakes were applied, but always returned to normal retraction. When the firemen set up the aerial to operate the ladder, it would work in a normal manner, without any problems. The manufacturer's representative could not decide or find a solution at that time, other than to send the aerial back to the factory. The representative then told the chief to tie a rope around the bottom rungs of the ladder sections during travel, in the event the aerial had to be used before sending it back to the factory. He advised to remove the rope after arriving at their destination if the aerial ladder had to be used. The fire department mechanic was not satisfied with this report, so he called for another opinion. I was working at that time for an independent "Third Party" x-ray and aerial apparatus testing company, and I was called to check the aerial before they sent it back to the factory. It took less than an hour to solve the problem, and the apparatus was put back in service.

What caused this apparatus malfunction?

The manufacturer's name Maximum should indicate what kind of an aerial ladder it was.

1. Maxim 1970's - 100 Ft. Aerial Ladder:

Southern California Answer:

This aerial ladder is made of stainless steel, and the rollers for the extension sections were made of a canvas material called phenolic. The ladder sections were extended by a hydraulic cylinder and cables. When the apparatus was driven and brakes applied, the tip and mid extension sections would extend a considerable distance, and then quickly retract back, according to speed and stopping force. Since this was a hydraulic cylinder extended ladder, there would have to be a pressure relief valve in the hydraulic system. Here is what happened. The weight of the ladder sections, with the momentum of stopping the apparatus, created a surge of weight that the pressure relief valve could not resist. Thus, the force would let the hydraulic pressure bypass, causing the aerial sections to extend until the pressure equaled. The pressure valve was checked, and the pressure setting was lower than the manufacturer recommendation. The spring had weakened in the valve; and after adjusting the relief valve pressure a small amount, according to the manufacturer's recommendation, the problem went away and never came back during the lifetime of the aerial.

2. Grove 85' Mid-Ship Chain and Cable Extension - Aerial Ladder:

Illinois:

A factory representative was sent to this fire department to check out a problem they had with their 1970's chain and cable extended medium duty steel aerial ladder. When this ladder was extended fully at any elevation, the ladder sections would always twist to the right side, approximately twelve inches at the tip of the fully extended fly section. It didn't matter what elevation the ladder was; and after working for about an hour of trial and error, the factory representative advised the mechanic that the mid-section was twisted, the aerial should be removed from service, and the ladder section should be replaced.

The extended sections would always twist to the same side when fully extended. The aerial had never been in an accident. Before the aerial was returned to the factory, I was asked by the manufacturer to verify this problem. I was working in the area, so I could get to the location quickly. PTS was also working as a "Third Party" acceptance company with this manufacturer. After inspection and operating the aerial, and explaining my findings to the manufacturer, the problem was corrected. The unit remained in service and did not have to be returned to the factory. After a minor adjustment, the twist disappeared.

What made this ladder twist?

2. Grove 85' Mid-Ship <u>Chain</u> and <u>Cable</u> Extension -Aerial Ladder:

Illinois Answer:

Sometimes on chain and cable extended ladder sections, the extension chains get out of adjustment, causing stress on the ladder sections at full extension, or full retraction. Usually, there are two sets of extension chains on the base section, and cables on the extension sections. When chains are out of adjustment by one or two links, one of the chains will hit the extension stop before the other. The opposite side will continue to extend the number of links that are out of adjustment; and when the opposite chain hits the stop, the ladder sections will twist to the side. The more links the chains are off, the greater the sections will twist; and sometimes, the chains will break under stress or other problems will arise. On this aerial when the ladder sections were fully retracted, the left side rail bottomed on the stop before the right side. The right-side extension chain had slipped two chain links off the sprocket, and when the sections were retracted, the left side hit the retraction stop first. The chain and sprockets were both adjusted, so the back of both side rails hit the stops at the same time. Thus, the ladder sections had no twist at full extension; and the ladder was returned to service.

3. 50' Tele-Squrt:

Illinois:

This aerial was set up for operation. When the inner boom was extended, a very loud screeching noise would emit from the booms. It sounded like the guide blocks were worn and metal was rubbing somewhere inside the main boom section. The top mounted aluminum ladder was also checked for damage and scraping. After installing lubrication to the guide support blocks inside the booms, the noise was still there. This was not the only aerial where this problem had occurred.

What caused the screeching noise when extending or retracting?

This problem is one of the idiosyncrasies that were never foreseen and was a little more difficult to resolve.

3. 50' Tele-Squrt:

Illinois Answer:

This tele-squrt and many others would make a very loud screeching sound when the extension boom was extended or retracted. It sounded as if the boom bearing material was making this sound. Bearings were lubricated, but to no avail. The sound was very persistent and would not go away. The only thing left to check was the telescopic water line seals. The noise was coming from the dry seals. When the boom was extended, the dry seals would rub the anodized water line tubing; and if the water line tubing was also dry, it would create a screeching noise while extending and retracting the boom. The anodized tubing on the water line was wiped down with ATF transmission fluid as it is both a cleaner and a lubricant for the seals. The boom sections were then extended and retracted, and the noise ended. The excess fluid was then wiped away. Oil attracts dirt and grime, and ATF seemed to be a better lubricant for wiping the water line tubing. The practice of wiping down the anodized water lines on all pre-piped waterways with transmission fluid has proven to cut down on maintenance and noise.

4. 1970's - 100' Telescopic Ladder:

Colorado:

PTS was called to give a "Third Party" opinion concerning the operation of this department's 100 -foot aerial ladder truck, before they send it back to the manufacturer in case there were other areas to consider. The fire department mechanic could not understand why this one-hundred-foot aerial ladder was malfunctioning when the ladder sections were extended.

This one-hundred-foot aerial ladder was set up for operation, then elevated to sixty degrees, the elevation lock valve was then applied. The ladder sections while extending had excessive side jerking, and the extension were very slow and erratic. The extension was acting as if one of the mid-sections was bent excessively, or if the guides were worn or binding, of if there was air in the hydraulic system, or if the pump was malfunctioning. This would be a very unsafe ladder to use in any emergency. The aerial ladder sections were retracted and lowered to a horizontal position and slowly extended over the rear of the chassis for inspection. The aerial and chassis had never been in an accident and was not operated every day. During the inspection, there were no signs of damage to any of the sections, rollers, cables, or guides. Cables were all in adjustment. Hydraulic oil was clean and full. The hydraulic pump was operating normally. The problem with the erratic extension was found. Showing the mechanic and operators the problem, after working with the unit for about three hours, the aerial apparatus was put back into service, and was operating like a new unit with no additional problems.

4. 1970's - 100' Telescopic Ladder:

Colorado Answer:

This one-hundred-foot aerial ladder would barely extend, would jerk, and twist side to side. As previously stated, there was no physical evidence of damage to any of the ladder sections or cables. There was an excess of hard, dirty, dried Lubriplate grease on all the ladder section guides. It appeared that the ladder sections had never been cleaned for a long time. In the early years of aerial design, there were not many choices of lubricating greases. Lubriplate was the most recommended grease since the 1950's but tends to harden with dirt and dust. Another problem there was an excessive amount of grease on areas that were not guides. By trial and error using different lubricants, PTS found that Silicone grease lasted longer, was easier on the guides, and did not harden with dirt and dust. At the auto parts store, the firemen purchased six spray cans of brake cleaner which evaporates and leaves no residue on the ladder sections, a few rags were needed to wipe off the excess grease to clean the rails. It took less than an hour to spray and clean all ladder sections. Then Silicone grease was applied very lightly to only the guides of the sections. The aerial ladder was retracted into the travel position, then raised to 60 degrees elevation, locks applied, sections extended smoothly without chatter or resistance, and showed no twist or damage. No other action was necessary. This condition was found on many other ladders; and when they all were cleaned and properly lubricated, they all extended effortlessly as this aerial did. Proper maintenance will extend longevity to equipment.

5. 1970's - Tractor-Drawn 100' Aerial with Mechanical Outriggers:

Massachusetts:

It was late one night when delivery from the factory of this new tractor drawn one-hundred-foot aerial ladder arrived at the city. The fire department personnel were informed by the delivery person on operating procedures, without operating the aerial as it was late. He explained how to jackknife the apparatus, extend the manual jacks, place the pads on the ground, and manually screw down the outrigger feet. It was a day or two after the representative left, when the apparatus was set up for routine operation by the department firefighter. The aerial ladder was lifted out of the bed and raised to sixty degrees. When the sections were fully extended, the tip of the fly section would twist toward the side rotated. The Chief called back the manufacturer representative, and his answer was the sections were bent and the aerial should be kept out of service, and he would come back later to check it out. PTS was called for a "Third Party" inspection to check the apparatus for ladder damage, and any possible additional areas for the manufacturer to check. I had the operator set up the unit as the representative had explained to the department personal. After observing the set-up of the aerial, and checking for any other problems, the apparatus was immediately put back into service, and the ladder sections showed no twist, and were extending and operating the way they should.

Why did the ladder sections twist from side to side while they were operating it?

Similar circumstance

Picture courtesy of John E. Hinant

5. 1970's Tractor-Drawn 100' Aerial with Mechanical Outriggers:

Massachusetts Answer:

When this tractor drawn aerial ladder was delivered to the fire department, the instructions for setup was to jackknife the tractor, extend the manual outriggers to full extension, and lower the screw jacks about <u>two inches</u> from the ground plates. The fire department used these instructions to set up the apparatus. When the extended ladder sections were elevated and rotated, the ladder and chassis weight would shift and lower the outrigger to the ground; therefore, the tip of the fly section would twist excessively toward that side. We then bedded the ladder, and started again, setting the outriggers firmly to the ground. After the outriggers were repositioned, the aerial was elevated, as before, and rotated without any twisting of the fly section. It is possible that the manufacturer's delivery representative had the setup information misconstrued. The firefighter had no idea, as they were only following department SOP. It is very difficult to set up an aerial with manual outriggers, especially when the terrain is uneven. Extreme care must always be taken when operating any aerial on irregular terrain.

6. Articulating Snorkel / Aerochief / Calavar / Bronto / Simon:

Virginia, Maryland, Florida, N.Y., Canada:

When the upper platform boom is raised, and then the lower boom was raised to or beyond **vertical**, the platform would travel an excessive free upward distance, when the lower boom reached or passed ninety degrees **vertical.** Bringing the lower boom back to the storing position, after passing the ninety-degree **vertical**, the platform would free fall a great distance, the opposite direction. The fire department was told by a manufacturer service representative, that the boom hinge pins and bushings were excessively worn, and had too much play, and all pins and bushings needed to be replaced.

The aerial did need to be taken out of service, but was this the correct assumption?

6. Articulating Snorkel / Aerochief / Calavar /Bronto / Simon:

Virginia, Maryland, Florida, N.Y., Canada Answer:
There are various sizes of turntable bearings to support and rotate different makes of aerial platforms. One of the most destructive means for turntable bearings, is over the road travel shock. The booms are pressured down onto the travel supports, to keep the booms from bouncing up and down on the support, and that puts reverse pressure on the turntable bearing. As the apparatus travels down the road, the vibration of the boom pounding downward on the travel support sends a shock back to the turntable bearing. This will wear the inside ball races and put excessive indentions into them from the ball bearings inside the bearing race. Usually, the most wear is in the front or rear of the bearing. It is very important to know the allowable up and downplay between turntable bearing races. The wear is measured with dial indicators in thousandths of an inch. When there is excessive wear, or up and downplay in the turntable bearing; and when the lower boom is raised vertical or past vertical, having the upper boom unfolded, the platform can drop or raise in a free fall way, according to the amount of wear in the bearing race. Excessive amount of bearing race internal wear could cause injury. It is very important to keep an eye on the wear of the turntable bearing, especially if there is excessive mileage on the chassis. Manufacturers have specifications on the amount of bearing play they recommend, and this should be inspected frequently.

7. 1980's - 100' Steel Aerial Ladder:

Wisconsin:
The Quality Control Inspector at one of the larger aerial manufacturers was performing a final acceptance and load test for a new one-hundred-foot aerial ladder, at the plant test area. He was following the NFPA procedure manual and measuring deflection as noted in the pamphlet. There was a steel cable attached from the tip of the extended fly ladder section rung, to a large metal tank filled with water hanging in the air about 3-feet off the ground. The QC inspector was using a yard stick to measure deflection and checking any lowering of the tank to the ground. PTS was at this factory to accept a repaired aerial for another fire department and was close to this QC in-house inspector. He stated to me that the factory had replaced the ladder sections to this aerial with new manufactured sections, and he couldn't understand why these new ladder sections keep failing the

load test. He said this was the second set of new ladder sections that had failed his test. The ladder was elevated to sixty degrees, fully extended, rung locks and elevation lock applied, and water weight hung by steel cable above the ground about three feet. He had marked the elevation cylinders with a marking pen on the shafts next to the seal, to check for any drifting. The marks were still visible after an hour had passed, yet the water tank lowered more than was acceptable. The QC Inspector had condemned the ladder sections, until I showed him what the real problem was.

Why did the QC inspector state the ladder sections kept failing? This situation was explained earlier in this book.

7. 1980's - 100' Steel Aerial Ladder:

Wisconsin Answer:
The factory QC inspector was performing the final test of a brand-new light- duty one-hundred-foot steel aerial ladder. His test procedures were following the NFPA standards and manufacturer's recommendations. The inspector had the ladder fully extended and elevated to sixty degrees, with rung locks applied, and weight hanging off the ground. He was measuring the bottom of the weight tank with a yardstick, as shown in the NFPA pamphlet. For the cylinder drift down test, the inspector marked the lift cylinder shaft at the seal with a marking pen, after the aerial was set and positioned for testing. The standard was to mark the cylinder shaft then wait for one hour to see if the mark remained. If the mark was still visible, the test was OK. I asked the in-house inspector if he was familiar with dial indicators. The inspector did not know how to calculate drift down. He only used a mark on the cylinder shaft. When he marked the cylinder shaft with a marking pen, he made a one-eighth inch line. This calculates to one hundred twenty-five thousandths of an inch. If the cylinder shaft drifts inward one and one-half thousandths per minute or ninety thousandths per hour, the inspector will still be able to see the mark on the cylinder shaft. With only one and one-half thousandths of cylinder drift per minute, the tip of the one-hundred-foot aerial ladder will lower seven and one-half inches in an hour. He was not able to see the small difference in the mark on the cylinder shaft, so the inspector determined the sections were bending. The QC inspector changed ladder sections three times before I showed up. After the QC inspector saw how I tested the aerial ladder with dial indicators, he told me later that the factory had changed their testing procedures.

When I helped develop "Third Party" inspections, supplementary ways to help determine wear or discover internal problems was added to the regular inspection procedures the NFPA recommended. The most effective way to determine if ladder sections are bending or if any cylinder in the system is malfunctioning, or has internal drifting is to apply additional techniques. There is a formula to calculate how much the tip of an aerial ladder fully extended and raised to any elevation, will vertically lower when the lift cylinder has minor internal leakage. A cylinder drift down test using dial indicators is performed. It has been proven that the drifting is consistent by the minute. The only variations are when there may be a weight change on the cylinder pressure while testing. After five minutes, the drifting does not change all the way to sixty minutes. You will be able to calculate how far the tip of an aerial ladder or aerial ladder platform lowers, usually after the five-minute test. Since using dial indicators in testing apparatus, manufacturers have set acceptable allowances for cylinder drifting. It's amazing how this small tool can tell such a large story.

8. 1980's - 100' New Aerial Ladder Platform:

Delaware:

PTS had worked with this apparatus manufacturer as a "Third Party" acceptance company for several years, and I was very familiar with the aerial platform apparatus about to be delivered to this fire department. There was another "Third Party" Inspection Company working in-house with the manufacturer currently. A new aerial platform with over one hundred feet in length was to be delivered, and the final acceptance was to be performed by PTS at the fire department facility. This aerial had been previously tested and certified at the factory by another in-house "Third Party" Inspection Company. This fire department was one of PTS's long-time customers. I advised the fire chief prior to delivery, that if this aerial outrigger structure was not updated for the new longer designed ladder sections, it would probably fail the load test. When the manufacturer heard what the chief had been advised, factory representatives were also present when the unit was delivered, and those representatives were also advised what the chief was told. To eliminate any confusion, I ask the factory representatives to set up the aerial for testing, per their standards. They were to operate the aerial, and to follow the instructions I gave for the load stability test. The aerial platform was set up for testing, and sections were extended and raised above horizontal to clear all chassis objects. The aerial was a rear mounted aerial, so the full extension was over the front of the vehicle, The load weights were hung below the platform and attached in a manner that if the aerial would become unstable in any position while rotating three hundred and sixty degrees, the load would release before any damage to the aerial platform could result. The aerial platform did fail the load test at forty-five degrees off each side front and rear of the chassis when the ladder sections were rotated over the sides at full extension. The representatives called the factory to confirm the platform did fail.

Why did the aerial platform fail the stability test, when it had been previously tested at the factory?

8. 1980's - 100' New Aerial Ladder Platform:

Delaware Answer:

The 1980's were the years of newly designed aerial ladders and aerial ladder platforms. Manufacturers were changing over to a heavier duty aerial device that would support a larger capacity, higher reach, and have pre-piped telescopic waterlines. Since the early 1980's, PTS has performed in-house final inspections on apparatus for many manufacturers. This inspection was to accept delivery of the apparatus at the department headquarters. In advance, I advised the Fire Chief that if this new longer-extending aerial ladder platform sections did not have the proper outrigger spread distance for this size apparatus, it would probably fail the load test. There were manufacturer's representatives present at the fire department headquarters, to observe PTS performing the acceptance and load test. To make sure the manufacturer agreed to all the procedures, I asked one of the service representatives to operate the aerial ladder functions while I tested it for acceptance, as per the NFPA recommendations. The aerial ladder platform was set up by the representative as per their SOP. I checked the outrigger spread, and it was rated for a shorter ladder. I showed the narrow width outriggers to the chief, and the representative stated the aerial ladder platform had already been tested by another "Third Party" inspection company and **was certified**. The aerial ladder platform was raised from the travel support to clear all objects around the chassis. The operator was then told to fully extend the sections and keep them in line with the chassis to start the test. over the rear of the chassis. The one and one-half to one test load was then attached below the platform (not in the platform) and supported slightly above the ground. The reason for keeping the load close to the ground and not loaded directly into the platform was that if there were any problems during the rotation, with the load close to the ground, it would release immediately, and take off any stress from the platform and chassis, to prevent a tipping situation. If the load was placed inside the platform, the unit would become unstable and tip over, at maximum reach. The test was performed as per NFPA and OSHA regulations, with the aerial platform in the most extreme position. When the test load was applied, the operator was told to start rotating, and the aerial platform became unstable at the forty-five-degree angle off each side front and rear sides of the chassis, and the load would drift to the ground, also causing the opposite side outriggers to raise off the ground. The operator tried to lift the load off the ground, but all that happened was the opposite side outriggers continued to rise higher off the ground. "The aerial ladder platform load stability test failed. Still, the manufacturer's in-house "Third Party" inspectors certified this unit as safe and stable." The chief was advised and made his final decision to accept the apparatus as is with repairs to come. **No** certificate of compliance was issued by PTS at this inspection, but the chief accepted the apparatus knowing what could happen. The outriggers were eventually modified to accept the proper load test and made the platform safe in any position of operation. If the in-house "Third Party" inspector had placed the test load in the platform, and raised the platform above <u>forty-five degrees elevation</u>, at full extension, when they rotated the platform, the reach would be shorter, and the aerial could pass at that elevation, and the old-style outriggers would work?

9. 1980's - 100' New Aerial Ladders:

Pennsylvania, L.I. New York This was the period in history when the change over from light duty aerials to a more heavy-duty capacity aerial was on the drawing boards at most of the aerial apparatus manufacturers. This was one of a "newly designed" heavy duty one-hundred-foot aerial ladders. This new aerial was set up for inspection in a large parking lot. The base ladder was raised above the travel support to clear any compartment mounted objects. The ladder sections were fully extended horizontally for routine inspection over the rear of chassis. The rotation manual disc brake lock was applied, and braking was also automatic through a rotation gear box. There was a slight wind blowing, less than five miles an hour. Suddenly, the ladder sections started to rotate freely without an operator at the control station. One could touch the extended ladder sections at the tip of the fly section and rotate the ladder manually around the chassis. This situation was found by PTS on more than one unit out in the field. This manufacturer has in-house QC inspection, and this problem was not noticed at the manufacturer. Unfortunately, this was the same manufacturer that did not understand cylinder drift-down as stated previously.

Why did the ladder rotate freely in the wind? This will be a tough question unless you are familiar with this manufacturer.

9. 1980's - 100' New Aerial Ladders:

Pennsylvania, L.I., New York Answer:

As the newly designed heavy duty aerial ladders and aerial ladder platforms were in the process of replacing the lighter duty aerials, there was one manufacturer who was upgrading their light duty aerial ladders to make it a more heavy-duty style aerial ladder. The problem was they continued to use the same light duty aerial ladder rotation motor and gear box assembly, with a manual disc brake control. The weight of the new ladder sections, extended with the wind blowing against them was too much for the light duty rotation gears and for the manual brake to support, so the sections could rotate even if the light duty manual brake was applied or not. The manufacturer was advised of this problem, and they reengineered. A new style heavy duty rotation gear box assembly was then installed on all the existing and new manufactured aerial ladders and the problem was corrected. I explained to the fire department how easy it would be for the elevated ladder to freewheel and drift into high voltage electrical wires. When rotating the elevated ladder sections, and releasing the rotation control, the aerial would continue to rotate. This problem was caught in the early stage of manufacture, but it could have been a serious situation. The in-house quality control, plus another "Third Party" inspection company missed this feature.

10. 1970's + Aerial Platform with Dual Elevating Cylinders:

New York, Virginia, U.S.

This aerial platform was set up for operation, the main boom was raised from the travel support, and the platform would quickly drift about one to two feet to the left and sometimes to the right. The inner booms did not have to be extended for this to happen. The main boom would stay in this mode until maximum elevation was reached. At the maximum elevation, the main boom drift would straighten out when hydraulic pressure was applied. When lowering the platform back to the travel support, the main boom would quickly drift to the reverse side; and after it was stowed on the travel support, the boom would straighten. The extension of the inner boom or booms did not have any effect on the amount of side drift but would only add side distance when extended.

What would cause the main boom to drift from side to side? This aerial platform is manufactured with dual elevation cylinders.

10. 1970's + Aerial Platform with Dual Elevating Cylinders:

New York, Virginia Answer:

Aerial platforms, with dual cylinders, have locking valves that require equal pressure. Some-times the internal lock valve pressure spring will weaken. When this happens, the aerial platforms will move sharply to the left or to the right when elevated and will stay in that position until full elevation is achieved. After maximum extension of both elevation cylinders, the main boom will shift back to a straight position. Inspectors assumed that the main boom hinge pin and cylinder pin bushings were excessively worn and needed replacing. When one cylinder relief valve spring gets worn or fatigued, that creates a different pressure between the relief valves on the cylinders. When the boom is elevated, the cylinder lock valve that requires less pressure to release will extend first. After the extension pressure equals with the other cylinder lock valve pressure, both cylinders will extend simultaneously, but the main boom will already be shifted to one side. When the main boom reaches full elevation, the other cylinder will fully extend, and the boom will straighten out. When lowering, the cylinder with the least lock valve resistance will start retracting first, causing the boom to shift until pressure equals when the boom is bedded in the travel support.

Aerial ladders and aerial platforms that have dual lift cylinders. can get out of sequence because of the holding valve release pressure settings are not equal. When raising and lowering the main boom and side movement or twisting is noticed, check the relief valves for equal pressure. Sometimes, stiffer valve springs must be installed into the holding valves to equalize their pressures.

11. 1970/1980 - 70'-90' Articulating Aerial Platform Setup:

Maryland, Georgia:
Articulating aerial platforms have been set up for operation on dry cement, asphalt, or on a hard surface, prior to operation. After operating the aerial, the booms were lowered into the travel support and locked down. There are usually two sets of **A-frame** outriggers, fore and aft of the chassis body which support the aerials for stability when in operation. After the booms are stowed, there is a hydraulic transfer valve mounted near the travel support that changes the direction of fluid from the upper controls to the outriggers valve automatically. This transfer will permit the operator to raise the outriggers for travel only after all booms are stored. There are swivel pads attached to the outrigger extension housings. Sometimes the outriggers are lowered for boom operation and set up on a hard surface. After the boom operation is concluded and the booms are stored for travel and the outrigger valve was actuated, the outriggers would not retract for storage as they are wedged into the hard ground surface.

What can the operator do to retract the outriggers so the apparatus can prepare for travel?

11. 1970/1980 - 75' & 85' Articulating Aerial Platforms:

Maryland, Georgia Answer:
Many of these aerial older articulating platforms have two sets of A-frame telescopic outriggers. When the outriggers are set up on a hard surface, like cement or asphalt, and the outriggers are pressured down past over extension, sometimes they will not retract

when the outrigger controls are actuated. Most of these aerials have a boom to outrigger transfer switch or valve mounted on the travel support. This switch or valve tells the outrigger control valve when the lower boom has been placed in the travel support, and that it is OK to raise the outriggers. Some aerial platforms have a <u>manual</u> transfer valve at the outrigger control station. There are times when the operator must fool the outrigger valve by gently raising the lower boom, and slightly rotating the booms to the right or left of the chassis, but not to rotate over the side of the chassis. This will shift the boom weight on the chassis from the opposite side. When this is done, the switch on the travel support can be manually depressed and held. This fools the outrigger valve to think the booms are stowed for travel. At the lower outrigger controls, slightly raise the outriggers (break loose) opposite of the boom rotation. Reset the boom back into the travel support and the outriggers should retract to storage road position and ready for travel. Do not raise the outriggers only loosen them from the surface, in case the boom gets rotated over that side, so they will still be very close to the ground. This is the quickest and fastest way to let the outriggers retract, but caution and care is always advised.

12. 1950's - 65' Mid-Ship Open Cab Aerial Ladder:

Maryland:

This fire department was in process of reinstating an old reserve open cab sixty-foot aerial ladder that had been removed from front line service and was stored outside in the weather for a long period of time. The fire department decided to use this aerial again as a reserve backup unit. The mechanic started the engine, engaged the power take off, set the manual ground jacks, and proceeded to operate the aerial. The ladder would try to raise but barely raised above the travel support, even when the engine rpm was increased. The mechanic shut down the unit and told the chief the hydraulic pump was malfunctioning and needed replacing.

The chief called PTS for a "Third Party" opinion and inspection, before replacing the hydraulic pump, to check out the aerial to see if this was the actual problem, due to the age of the aerial.

12. 1950's - 65' Mid-Ship Open Cab Aerial Ladder:

Maryland Answer:

This 50's open cab sixty-five-foot aerial ladder was stored outside in bad weather conditions for many years. When the hydraulic reservoir was checked for oil, there was a big surprise. The reservoir was full to the top of the filler, but the oil had thickened, due to years of condensation, to the point of thick mush which plugged up the pump and filters. The hydraulic system was flushed, new filters replaced, and ladder sections were properly cleaned and lubricated with lithium grease. The aerial ladder sections had no visible damage; and when operated after the servicing, the ladder functioned like a brand new sixty-five-foot aerial ladder. It elevated and rotated with very little effort and was put back into reserve status.

13. Aerial Ladder – Aerial Platform Height:

Have you ever checked the aerial ladder or aerial ladder platform in your department to confirm that it would reach the advertised height for which it was designed, or did you just take it for granted the height was correct? What would be the best way to figure out the height? Aerial ladders and aerial platforms are designed differently, so the tip rung of an aerial ladder is the maximum point to measure from the ground; and on an aerial platform, it is usually measured from the platform floor to the ground, and then you need to add an additional five feet for the operator to reach the maximum height. There have been aerials that did not meet their 'advertised' heights. There was a sign stating height of a special platform that did not coincide with the actual measurements.

13. Aerial Ladder – Aerial Platform Height:

ANSWER:
The only aerial platforms that elevate to a true working angle of ninety degrees vertical are the articulating type. Aerial ladders and aerial ladder platforms generally work from sixty to seventy-five-degree elevation. It would be very difficult to climb an aerial ladder that is ninety degree's vertical elevation, especially with sections fully extended. That is why aerial ladders are designed for a comfortable safe working and climbing angle.

There are different ways to measure maximum heights. On aerial ladders, attach a measuring tape to the fly section tip rung, elevate to maximum working degree of elevation, and then extend all sections to maximum extension. Measure the distance to the ground, and that will give the maximum height of the aerial ladder.

On aerial ladder platforms, simply attach a measuring tape to the bottom center of the platform, elevate to maximum height, and then extend all sections to full extension. Measure from the bottom of the platform to the ground; and on aerial ladder platforms, add five feet to the measurement. This will give the maximum reach height for the operator. There was an incident where an aerial ladder platform was advertised one hundred feet plus on the apparatus; but realistically, it only measured about eighty-seven feet to the bottom of the platform at maximum elevation and full extension. With an operator, the maximum reach only would be approximately ninety-two feet. If the ladder sections would have been longer to make up the distance, the outrigger system may have not supported the extra length. There is a 10' attic ladder mounted close to the platform on the ladder section tip; and at that time, the manufacturer's representative advised the fire department to use the attic ladder inside the platform so over one hundred feet could be reached.

Maximum advertised height of aerial ladders and ladder platforms does not represent the reach of the aerial. Aerial ladders are usually operated at an angle that an operator or fireman can climb with ease. This angle would vary from "o" degrees to approximately "60: degrees. These climbing angles will cut down considerably from the advertised height of the aerial but give side reach.

Articulating aerial platforms can reach a true ninety-degree vertical position and are a little more difficult to measure with a hanging tape measure. Still measure from the bottom of the platform and then add five feet for the average operator reaching height. There is another way to calculate boom heights, with measurements from hinge pins on the booms, hinge pin to bottom of platform, and turret hinge pin to the ground.

RESEARCH:

Aerial ladders and aerial platforms improved drastically since the early 1900's. Wood aerials from the early 1900's changing into metal aerials. The need to support these aerials when rotating the ladder and reaching over the sides of the apparatus was a new priority. Outriggers were now on the drawing boards. Manual spring lockouts were one of the first to help support the chassis. To help support the ladder while reaching over the side of the apparatus, telescopic steel housings were installed with a manual pullout steel tubing to reach beyond the side of the body with manual adjustable screw down supports with pads attached to the bottom of the screws. Years later when hydraulics entered the picture, this started the manufacture of various styles of outriggers to support all aerial apparatus. At first, hydraulic cylinders were vertically attached to the chassis frame under the turntable with manual shut off valves to control flow. These cylinders did give some chassis support from flexing, but still limited the aerial side reach. As time passed, longer reach and heavier new style ladders were being manufactured. New innovations of articulating aerial platforms were being introduced to firefighting. Outriggers to support these aerials had to be developed. A-frame style telescopic hydraulic outriggers seemed to satisfy the new articulating aerial platforms, usually four outriggers were used. Two at the front of the body and two at rear of the chassis. These types of outriggers have a double safety, as the inner housing would wedge into the edge of the outer housing when extended and they would also have a hydraulic locking valve. One aerial platform uses vertical mounted outriggers on all corners of the chassis, with an over center outrigger, that folds down at the front of the turntable and is connected to the frame turret assembly.

On aerial ladders and aerial ladder platforms, there are various ways to support the apparatus. One manufacturer had vertical outrigger cylinders at both sides of the rear chassis, with manual swing down cylinder under the front of the chassis, along with two telescopic H-frame outriggers connected to the turntable mainframe in the center of the chassis. Another manufacturer had four-fold out outriggers connected to the chassis mainframe, at the front of the body and at the rear of the body. Most heavy-duty aerial ladders and aerial ladder platforms have telescopic H-frame outriggers connected to a chassis mainframe to support the aerial plus use the chassis for counterweight support.

There is a manufacturer that builds a complete torque box chassis with four telescopic a-frame outriggers to support the apparatus in such a way that reduces excessive side extension of the outriggers.

Setting up the apparatus for operation varies with the multiple types of outriggers and apparatus. Many A-frame outriggers are extended to a stable position with tires still on the ground. Many times, operators overextend the outriggers to maximum extension, lifting the chassis and tires off the ground. Some units have greater stability when only the bulge is taken out of the tires, leaving about eighty percent of the chassis weight on the ground, and the chassis works as a counterweight. Manufacturers have a procedure for setting outriggers for their apparatus, and the operator should follow the instructions to the tee every time the aerial is raised. Short jacking an outrigger does not allow any aerial to have the stability required for proper operation when there is limited space for set up. Short jacking an outrigger is dangerous and should never be permitted. Sometimes the operator forgets he has short jacked one side of the chassis and rotates over that side, causing the aerial to fail.

Two of the most puzzling and complicated problems with firefighting are extinguish and rescue, especially when a two-story office building became a monster skyscraper towering over the downtown metropolis. Both have separate venues with engineering and have had centuries of trial and error, and both still exist without total perfection.

There was a thought of using helicopters with a rescue platform attached, for very tall high-rise buildings rescue. This was a great idea. The problem was with <u>rising smoke and wind</u> restricts the vision for the pilot to set up for rescue. This idea is one of many which have been considered to help in high-rise rescues. Helicopters are used every day in extinguishing ground fires and rescue worldwide. The helicopters have many versatile uses that have not even been discovered.

Aerial platforms have reached heights exceeding ten plus stories for rescue and are also used as a water tower operation to extinguish fires. When in the rescue mode, some of the higher aerial platforms have a chute attached to the platform for lowering victims from the fire to the ground, in a quick manner. Once in the chute and dropping to the ground, the material of the chute slows down the descent for the victims to reach the ground without injury. Other platforms have ladders attached to all the boom sections with handrails, to aid the rescued victims while they ascend to safety. The larger and higher reach platforms may have difficulty setting up in restricted areas, as they require a large open area to extend the outriggers, and to unfold the boom sections. This could also restrict the reach and height distance for rescue and fire extinguishing. The overall height measured for aerial ladders and platforms is at the highest elevation and fullest extension they can obtain. A one-hundred-foot aerial ladder elevates to a maximum of between sixty to seventy-five degrees for ease of climbing. When the ladder must reach a rescue, the ladder is lowered until it will reach the object for rescue. This reduces the height the ladder can reach. The one-hundred-foot ladder may only be able to rescue at sixty feet. The same is for platforms, where the reach distance can be restricted, for the height needed. Extinguish and rescue at high-rise towers needs to be a high engineering priority. Evacuation in emergency situations is complicated and difficult to control. We need to keep our thinking caps on and keep trying to find solutions.

Many of the larger aerial ladders and aerial platforms cannot respond or set up in restricted narrow streets and alleys. A great engineering innovation is the tractor drawn aerial ladder. This apparatus can maneuver down the narrowest streets and turn corners on a dime. Equipped with a second driver at the rear tiller, allows this exceptional navigation of the apparatus. The only problem with this type of equipment is that it's restricted to aerial height of around one hundred to one hundred ten feet.

There was another innovation, which was dual steering on single axle apparatus. They were smaller units and could get into complex road areas. They maneuver like the tractor drawn units but have all wheel steering. Not all fire departments have this type of unique apparatus. One of the great designs of fire apparatus was the Squrt boom. By taking an engine with a pump and small reservoir tank, adding a telescopic aerial waterline boom at the rear of the chassis, with about fifty plus feet of side reach, makes this a very versatile apparatus. When the downtowns moved to the countryside and started the two- and three-bedroom communities, this aerial concept exceeded any other means for fighting fires. Highway accidents were easier and quicker to get to, due to the speed and size of the equipment. Fire departments send quints and sometimes aerials to car accidents and special emergencies because they are specially equipped with emergency equipment.

Airports also have a distinctive requirement for specialized fire apparatus. Crash rescue trucks can race down the runway at great speeds while spraying fluids. They also have an innovation of piercing and spraying fluid inside the burning planes.

GROUND LADDERS GENERAL:

Ground ladders are installed on Engines (pumpers), aerial apparatus, and require a series of inspections and tests to check for safety and durability. Since the 1950's, NFPA has recommended some type of inspection and testing of the ground ladders and are recommended annually, or every five years based on the fire departments priorities. "Third Party" inspections of aerial ladders, aerial ladder platforms, pumpers and aerial platforms could not be complete without including the ground ladders that accompany the firefighting apparatus

From early BC, some means of scaling heights to get from the ground to a higher position, has been an achievement of engineering. There were no drawing boards or ways to calculate strength or section modulus of materials, in the early days. I suppose they started with a tree limb, shortened the branches, and had a way to scale to higher elevations. Primitive as it was, it served the purpose at the time. There were vines and various materials weaved into ropes that could be used to lower or repel from, to get to a lower position. One could slide down from a height once the vine or rope was anchored but going from the ground to a higher level when the vine or rope was not anchored was a different situation. On a vine a knot was tied every few feet or inches. This would give at least a foothold, to prevent sliding down the vine and easier to climb. Then the rope ladder arrived with two ropes or vines, with horizontal foot straps anchored to the sides of the ropes or vines. This would not only give each hand a place to grab, but also a place for their feet, to ascend or descend. The rope or vine had to be anchored before this could happen. The single tree branch ladder and the rope ladder were then transformed into a dual vertical post, with horizontal branches tied to each side. At that time, no string or wire was available to attach the rungs to the side poles, so they possibly used strips from animal hides to attach the rungs to the side poles. Now, we have a means to reach heights in an easier way than ever before. Just rest the ladder on any object and one can climb to new heights. Ladders called "scalae," scaling, brooms, poles called "perticae," and even life nets, were used for firefighting as early as the fall of the Roman State. This was the first firefighting known to history, the first century BC. As centuries passed, ladders were used in wars for climbing walls, along with rescuing victims from heights, and fighting fires. Ladders were limited on how high they could reach, and this was based on the weight of the ladder and the length of a single section. Single section ladders were used by Indians that lived in high mountain dwellings, where ladders were used to climb from one dwelling to another. Each family made their ladder according to the height they had to climb.

In this country in the 1600's, the single section ladder was still used, along with pole ladders for knocking down walls. Ladder technical knowledge advanced greatly through the centuries. In the beginning, ground ladders used for firefighting in New York City were left outside. They were hung on fences so volunteer firemen and citizens would have easy access when an alarm came in. New York City purchased their first ladders with hooks mounted on the ladder tips in the early 1690's. Who manufactured these ladders at that time was not recorded, but most ladders came from Europe?

One of the most used and universal ladders was introduced in the mid 1860's from Elberfield, Germany. This would be the Pompier Ladder. The Pompier Ladder was developed and used as early as the 1820's, but fire departments did not show much interest in this ladder until the 1840's. Other models of the Pompier were invented, and they were called Scaling Ladders. When heights reached above a normal ground ladder reach was needed, Pompier Ladders were then used in conjunction with ground ladders to reach windows or any opening that could not be reached with a normal ground ladder for rescue. This ladder had a single pole with a large hook on one end, and cross bars running through the main beam equally spaced for footing. As taller buildings emerged, the need for this type (Pompier) ladder was in great demand. In the mid 1870's, the Pompier ladder was introduced to St. Louis Fire Department, by an immigrant who came to America, he had worked previously for the German Manufacturing Company where the Pompier ladder was developed. It could be said that St. Louis Fire Department introduced this type of ladder to America. Years later, there was an American developer that changed the foot rungs of the Pompier Ladder into folding members, but the original design seemed to last forever. These ladders were used to scale large buildings from window to window and were used until apparatus mounted aerial ladders came into the picture.

Picture courtesy of Fire Engineering

Single wood ladders with two beams and hooks mounted on the tips of one end of the beams, started to appear around the 1870's. These wood ladders were used as climbing and then as roof ladders. So now the process has started, and the development of various types of wood ladders were now emerging into the New World. Records are not clear as to when ladders with more than one section arrived, but signs show it was close to the 1870's, or the early 1880's, when extension ladders were first manufactured. One of the main reasons was the single wood ladders could not reach very high on multi-story buildings, so many lives were lost because the single ground ladders were too short. Before multi-section ground ladders were developed, firemen would tie single ladders together to gain a higher reach, and many failed.

The demand now was for longer ladders to reach greater heights, and this would require multiple sections. At first, these ladders did not prove to be safe, and many failed when tested. When experimenting with the extra-long multiple extension ladders, many lives were lost. After continued experiments, it was determined that wood extension ladders should be no longer than sixty-five feet in extended length. Most wooden ground ladders were manufactured using Douglas Fir. This type of wood had the most strength and durability. The rungs, on the other hand, were made with White Ash. Wood ladders required serious maintenance to preserve their durability and strength. There were two types of ground ladders. The single beam ladder with rungs evenly spaced and mounted directly in the side beams. Truss ladders have two rails on each side, and three types of rung mountings. The rung mountings into the rails determine how the ladder would be positioned against the structure and climbed. The first is a truss ladder with blocks mounted between the beams, and the rungs mounted into the blocks. This type of truss ladder is placed with the truss side, away from the structure. The second and third truss ladders have blocks between the beams, and the rungs are mounted directly into the top or the bottom beam. This truss ladder is to be set up with the truss inward, and toward the structure for climbing. Wood ground ladders were very heavy, and a sixty-five-foot truss wood ladder could weigh up to two hundred ninety pounds. Today, there are still wood ladders in the fire service. One of the main reasons departments use wood ground ladders is because they have electrical resistance if the firemen accidentally raise them into overhead electrical wires.

In the early 1930's, two manufacturers of ground ladders introduced into the fire service a straight beam aluminum fire service ladder, and an aluminum truss ground ladder. Aluminum was much lighter than steel, more durable; and although the melting properties were very low, it remained the best material for the design of ground ladders. This also opened the door for a few manufacturers of aerial apparatus to begin manufacturing aluminum ground ladders with their own personal design. The ground ladders from the apparatus manufacturers finally faded away through the years; and today, the two original manufacturers of aluminum ground ladders continue to manufacture their style of ladders.

A new product was developed years later, not only did it insulate and resist electrical impulse; but when properly developed, it had dielectric properties that exceeded wood products. Here we have another innovation for ground ladders, and a safer material to prevent electrical interference. This would be the fiberglass ground ladders. Fiberglass has an extreme resistance to electrical charge. The only time a problem would void the resistance of the fiberglass or wood is when there is moisture. Fiberglass ground ladders have infiltrated the fire service with good response.

There is also a versatile multi-service fiberglass ground ladder that has many different forms of extension and folding. When they were first inspected and tested, per the NFPA requirements, the manufacturer recommended this type of ladder not be subjected to these stressful tests. Although these special ground ladders may not pass the rigid testing that is recommended by the NFPA, they are still used by many departments due to the versatility they possess.

TESTING WOOD GROUND LADDERS:

NFPA recommendations for ground ladder testing started around the 1950s. There were many fire departments with wood and combinations of wood and metal ground ladders. Individual apparatus and pumpers were recommended to carry a minimum footage of ground ladders, per the NFPA recommendation at that time. Fire departments were doing in-house inspection of their ground ladders. It wasn't util the 1980's when "Third Party" inspection companies were new on the block; and many of their test procedures were trial and error. It was the beginning of a new way for ground ladder inspection and testing. Wood ground ladders were tested the same as metal ladders. In the beginning, a tremendous amount of wood ladders was destroyed. Why did so many ladders fail? Was it the wood that was at fault, was it the inspector and his test equipment that destroyed these ladders, or was the test weight too much? There was no guarantee that when the wood ladder rails were assembled at the factory were matched sets. Many fire departments opted out of inspecting their wood ladders and only serviced them with varnish. On many of the wood ladders stress tests, one side rail would weaken much sooner than the other side. When this happened under the horizontal deflection test, weight would shift to that weaker side, and that rail would snap or crack. The weight would shift much quicker when the inspector used water or sandbags for test weight. When the rail starts to show cannot support the test weight, was the time to stop. There is no way to calculate the amount of shifted water weight that the rail supported before failure, or the amount of sand weight that was applied to a single rail. A water container holds up to 500 lbs. and was first used in horizontal testing of ground ladders

From experience with water weight, one of the very first PTS inspections was a thirty-five-foot ground ladder that broke during the load test. While lowering a water weight container onto a thirty-two-inch square support centered on the wood extension ladder, the weight shifted to the far rail, just as the maximum load weight was reached. When this happened, the overloaded rail quickly snapped, and the ladder was then scrap. That was the last ground ladder PTS ever tested with water weights. When load testing ground ladders, we started using equally distributed calibrated weights placed on a thirty-two-inch frame on the top center of the extended ladder. When using a dynamometer with calibrated weights, it is much easier to test ground ladders and to determine when failure could occur. Wood ladders are very similar to aluminum ladders, as they both seem to have stress load memory. Once the ladder has been stressed in the horizontal position to a certain load test, and the ladder passed the test, the same load can be applied later, and the wood or aluminum ladders will have no problem achieving that test, if no damage to the ladder has occurred since the previous inspection. After a close inspection of wood ladders, visually checking for splits or dark streaked areas in the rails, and whether they have been properly maintained with varnish and sealers, the ladder can now be set up for testing. Very few wooden ground ladders, thirty foot or less in length, have ever failed year after year, when properly maintained. Thirty foot and longer extension ladders must be tested with extreme care; and if there is any indication, they may not accept the load stress test, they should no longer be tested. Other wood ground ladders have shown indications where the possibility of failure could occur. These ladders also need to be stress tested with caution. Rung locks and rung torque tests 'usually' never fail their tests. The horizontal bending is the most destructive of all the tests, very few wall or single section wood ladders ever failed. Using a calibrated dynamometer when applying a stress load to an extended ladder placed in the horizontal position will indicate when a ladder will or will not support the stress test load. By watching the needle of the dynamometer as the test load is applied, when the needle stops moving before the correct amount of load weight is applied to the ladder, this will give an indication that the ladder cannot support any additional stress; and if you continue to apply additional weight, the ladder will fail. This doesn't mean this ladder was not a good ladder for service; it means the new tests required by the NFPA, should have been adjusted for this ladder, so it would not be subjected to this extreme test, when the required load capacities were not mandatory at the time this ladder was manufactured. This is how difficult it is to do a proper inspection and load test on wood ladders, even though we used calibrated distributed weights, equally placed in a holder, wood still has its own characteristics.

There was a time when the NFPA recommended loading the extension and single ladders for testing with bags of sand to achieve the proper load tests. Sometimes the bags of sand were placed on one rail more than the other, and the ladder would fail. After the sandbags were placed, there was no way to determine the distribution of weight or how much weight was placed on either ladder rail. This problem is the same when using water weights. There have been many failures of wood and aluminum ladders, due to the uneven sand and the water shifting onto one side rail. These failures were not necessarily caused by the ladder construction, but mostly by the inspector.

Wood and aluminum roof ladders were first manufactured with smaller diameter roof hooks, and these smaller diameter hooks would fail the updated NFPA roof hook load test. When testing these ladders, if the roof ladder didn't fail the horizontal bending and rung torque tests, but had the 'smaller' diameter hooks installed, the hooks could be replaced in the field, with new 'larger' diameter hooks furnished by the manufacturer, and after re-testing these new roof hooks, the ladder could remain in service.

The wood and aluminum and fiberglass 'collapsible, folding attic, or so-called closet' ladders have a lower stress test and usually have no problems, unless there was already damage to the rails, steps, or swivel feet. Just imagine how many centuries' wood ladders have been in fire service. Now, they are slowly fading out of service and being replaced with lighter durable aluminum and fiberglass ladders. I know of many fire departments who take pride in caring for their wood ladders, and it certainly would be shameful for anyone to ever test and damage their trophy ladders.

METAL GROUND LADDERS GENERAL:

Wood ground ladders and wood aerial ladders remained in service even after metal ladders were introduced in the early 1930's. Then, metal ladders began to replace the wood ladders. Major aerial ladder manufacturers began production of their own style and brand of metal ground ladders. Basically, in the beginning, there were two types of metal ground ladders manufactured with different types of material. They were solid beam and truss ladders, using steel, wood, or aluminum. Aerial ladder manufacturers continued making wood ground ladders. Steel ground ladders quickly faded away, and aluminum seemed to be the best material for manufacturing new, safer, lighter ladders. Many fire departments did not want metal ladders, due to the electrical conductivity. Wood ladders remained in service, even though the aerial ladder manufacturers were producing their own style of metal ground ladders.

There are five basic ground ladders used in fire service. The first is the straight ladder or so- called wall ladder. This ladder comes in various sizes up to twenty feet in length. The second is the extension ladder. These ladders range from fourteen feet to sixty feet in length, and may have multiple extension sections, using both rope halyard, and cable halyards. The longer extension ladders, usually forty feet in length and longer, are known as Bangor Ladders, and will have poles attached to the tip of the base ladder, for assisting firemen to elevate the ladder into position. The third ladder for fire service is the roof ladder. Basically, this is a wall ladder with two roof hooks attached to one end of the ladder beams. These ladders usually come in lengths from fourteen to twenty feet in length. Sometimes a shorter wall ladder is used for a roof ladder and two hooks are installed on one end. The fourth ladder is called attic, closet, folding, or collapsible. These ladders can come in lengths from six feet to around twelve feet in length. Some are still made of wood, but most are aluminum. These ladders are easy to maneuver, fold into a small width, and are easy to store. Their capacities are reduced, due to the way they are constructed, and they are designed for only one person to climb. The Pompier or Scaling Ladder is the fifth type, and slowly is fading from many fire department ladder requirements. This ladder was used to hook on windowsills and used to scale tall buildings from window ledge to window ledge before aerials were available. The single beam with foot rungs equally spaced and a large steel hook mounted at one end, made this ladder a very universal tool throughout centuries. Some or all the five ladders listed above may be found mounted on or stored under aerial apparatus compartments, on sides of pumpers, and on rescue units. These ladders are recommended for inspection and testing per NFPA current regulations. Some ladder trucks carried over two hundred feet of ground ladder complement, in addition to the aerial. There are many departments who have their ground ladders inspected and tested, along with the aerial inspections, on a yearly basis. A few departments postponed their testing up to five years and performed only visual inspections yearly. An excellent reason to inspect and test ground ladders more frequently, is the possibility of fire damage, structural damage from bad handling, and wear from the travel support or storage compartments. Usually, when fire damage or smoke shows on the rungs or rails, most departments will remove that ladder from service. This damage or indication is probably the easiest to notice, along with the roof ladder tips with rails bent inward from dropping on the hook ends. Sometimes, roof ladders are dropped during rescue or emergency situations and may not be noticed, so the ladder is kept in service. One would not imagine that travel supports do major damage to ground ladders; however, clamps locking the rungs to the travel support can cut the rung or wear an area down. The travel support with the ladder hanging down can wear into the rail sides. Steel ladder supports mounted on the side of pumpers or rescue units have this type of ladder mounting. There are supports that swing the ground ladders, which are stored flat on top of the body, down to the side of the chassis for easy removal. These travel supports may have edges that rub the ground ladders while traveling down the road. There are ground ladders stored in compartments, at the rear of the chassis body, with no locking devices for the ladders, so they could move about and rub objects inside the compartment, creating damage to the rails. Some ground ladders are stored where they rest on body mounting bolt heads, or close enough that they could sometimes rub and wear the ladder components. It is very difficult to manufacture a travel support or compartment that does not cause some sort of damage to the ground ladder in one way or another over a period. An excess amount of over-the-road travel, as stated in the "aerial ladder section" above, can also cause severe damage to ground ladders, just from being mounted to the travel supports.

TESTING METAL GROUND LADDERS:

Setting up ground ladder equipment for testing requires a 'level' ground area for the stands to sit, for the horizontal bending tests. Any sloping of the ground will result in an overload condition. When the extension ladder is fully extended, and the ladder pawls "dogs" are set in place, the pawls need to be checked for proper operation, wear, broken parts, and lubrication. There are times when the return springs will malfunction, or the pawls will jam inside the housings. When the test load is applied to rungs, check the locks, this will also check the rung strength, and the pawls for holding. Some of the earlier manufactured aluminum extension ladders, had support guides which wrapped around the base ladder and when extending the fly ladder, the guides could damage the inspector's hand, if the pawls did not set properly and the extension section happened to slide backwards. These ladders had to be extended with caution, even if they passed all tests, as they could still cause an accident. It's an accident that can and has happened. One must be very cautious when setting up any extension ladder.

The horizontal bending test method is used for all ladders: wall, roof, attic, and extension. This is the most critical ground ladder test; and if not done properly, **will damage** any ladder. I have tested the same fifty-foot Bangor ladders, year after year, with NO failures. The vertical deflection on the fully extended fifty-foot Bangor was close to three feet. The center of the extended ladder sections was very close to the ground, but not touching it, yet the ladder supported the total recommended test 500 lb. load.

50 ft. Aluminum Bangor Ground Ladder Picture courtesy of C. Dewey

Aluminum ladders inspected and tested seem to remember the load applied and test satisfactorily at each re-inspection. Applying the load equally to the rails is very important. There are ladders where one side of the rail could have had heat exposure and would be weaker than the other side rail. For this ladder, before any test load is applied, the special heat sensor should alert the inspector, and a special hardness test would confirm if there was rail weakness from heat. To verify if one side of the rail was weaker than the other side, the load should be applied very carefully and 'equally' to both rails. The gauge on the weights will indicate when the weakest rail cannot support additional weight. If the ladder has both side rails with heat damage, and the hardness test confirms material weakness; when applying the test weights and the load gauge indicator does not move anymore when applying the load, this is the maximum weight the rails can support; and if more weight is applied, the rails will fail. When the dynamometer indicator needle stops moving at a low weight, and the ladder sections continue to lower, the weight must be removed from the ladder, as the sections are about to fail, and the ladder should be taken out of fire service. Many ladders have failed while testing with water weights, because the water has shifted to the weaker side rail and puts this rail in an overload condition. This is how most extension ground ladders are destroyed. Extension ladders are very different from wall or single ladders, because the four rails are supported by guides. If there is a weak rail on any ladder section, and the test weight is a container filled with water; when the weak rail starts to deflect, the water shifts to the weak side, causing an overload condition, and the support guides will break apart. Another synopsis where failure to a ladder can happen quickly is when the ladders are set up for testing on a slope. When performing the horizontal bending test, using water weight, the water will shift to the sloped side, and most often the ladder fails the test. There is no way to prove after the failure, that the ladder wasn't defective, or unfit for service. Very few ladders fail the horizontal bending test if it is properly applied. The only ladders that could possibly fail are ladders that were visually found defective, and these ladders should be removed from service without further testing. Single ladders with slightly twisted rails can be tested, and usually pass the horizontal bending test, if the twist is minor. When the rails have a severe twist, it is recommended not to use the ladder for fire service. Many rails are twisted from dropping, or from some unnatural occurrence, or from improper travel support mounts on the apparatus.

Picture Courtesy C. Dewey

A fire chief in the southern part of Texas needed his ground ladders re-inspected and tested for insurance purposes. He said the prior year he had a "Third Party" Inspection Company come to the station to inspect and test all his ground ladders. After the inspectors were finished, they had damaged and broken most of this departments ground ladders. One of the inspectors told the chief the ladders that failed were defective. There was another twenty-four-foot extension ladder the inspectors did not test, as it was like the ladder they had already destroyed. The inspectors told the chief to remove this ladder from service, because the one identical to it had failed, and it would unquestionably fail. The "Third Party" Inspection Company told the chief to replace all the ladders that failed. I told the chief if the twenty-four-foot ladder did not have heat or structural damage, it should have passed all tests with flying colors. The chief of this fire department requested me to come to his fire station and prove my statement on the twenty-four-foot ground ladders. PTS went to the fire department. The longest extension ladder at this department was a new thirty-five-foot ladder, along with a couple of new twenty-four-foot extension ladders. The twenty-four-foot extension ladder the fire chief removed from service was chained to the firehouse wall. This twenty-four-foot ladder was taken down and was completely tested. The fire chief watched while the load tests were applied and removed. After the inspection, he stated, "Those SOBs broke my ladders, didn't they?" I'm sure there are many other fire departments that can relate to this same situation.

Roof ladders had a problem in the late 1970's and early 1980's, which was the existing 5/8-inch diameter roof hooks, originally installed by the manufacturer, would not pass the revised NFPA 500 lb. tip load requirement. When this 5/8-inch hook was found on a ladder in service, even though the ladder had no damage, bent rails or heat damage, and had passed all other tests, it was the recommendation to replace the smaller hooks with new 3/4-inch diameter hooks, furnished as a retro-kit by the aluminum ground ladder manufacturers. Each manufacturer would supply their new 3/4-inch diameter hooks. Several of the roof ladders had a rung or rail support missing at the tip of the ladder, where the hooks were attached. With the new hooks, a rail reinforcing member was also available to install. With this support installed, the ladder rails at the tip had a stronger reinforcing for the hooks. In the field, many roof ladders needed to be modified. When firemen or fire mechanics started replacing the new hooks on roof ladders in the field, there was a problem with the new style hooks. The hook sets came with different bends in the hooks, the sets got mixed and most hooks did not match upon installation. Some field installers didn't pay attention to matching the bend radius at the hooks tip, and the hooks got mismatched. When the hooks were tested, the hook with the deepest bend would hit the test block first, and the ladder would twist sideways, until the other hook engaged. A proper test could not be performed until the hooks were changed and both tips uniformly matched. This situation took considerable time to unwind and finally get roof ladders back into service without a hook problem.

14 Ft. Roof Ladder Hook Test Picture courtesy of C. Dewey

Wall ladders are one of the easiest ladders to inspect and test. Once it was determined there was no fire or heat damage, bent or damaged rungs or rails, the rung test and horizontal bending was all that had to be completed to certify this ladder to comply.

Attic, collapsible or closet ladders as they are called have a much lighter test load applied horizontally, and usually always passed. The most damage they have is swivel feet damage or worn and loose rivets in the swivel rungs. Very few have failed the test load, but the ten foot and longer ladders have much more deflection, and caution in testing is recommended.

Today, Pompier or Scaling ladders are few and far between on new aerials. When a truck mounted aerial is used, the complement of ground ladders on the apparatus has been reduced substantially. Small fire departments, that do not have an aerial ladder, and have only a few tall structures, usually have scaling ladders on the engines. To certify this ladder, the "Third Party" Inspection Company should check for damage or heat exposure, apply the test load to the rungs and hook, and issue a Certificate of Compliance.

Fiberglass ladders are relatively new to fire service and are inspected and tested like any other aluminum ground ladder. Fiberglass ground ladders also have outstanding durability, and a high electrical resistance. In the future, the fiberglass ladder will probably replace all wood ladders still in fire service.

Heat sensors became a new way to determine if heat has been applied to an area of the ground ladder. There is a small problem with heat sensors. They could give a false reading; and the sensor indicator could change color by hitting or rubbing it. This will indicate the ladder may have some type of heat damage. Heat sensors have expiration dates, and are supposed to be replaced by that date, even if the sensor has no indication of damage. When the sensors look O.K. and had no damage or heat exposure, but were expired, many fire departments that were previously told to change the sensors, did not, and left them on the ladders from year to year. If there is a manufacturer's record that shows the degree of the sensor's disintegrating, or how they became inaccurate, other than the expiration date, the fire departments should be informed. There can be six or more heat sensors on each ground ladder section.

In conjunction with the heat sensors, there are also information stickers that are required for ground ladders, such as, warning and ladder set up. Many ladders, while being stored on the apparatus, rub components, so the warning and information stickers also get worn or torn off. They usually don't get replaced until the annual inspection has been completed.

Good information makes it easier to follow instructions without having to make hard decisions."

EDIFICATION:

Some Part of the information below has previously been explained in the Maintenance Section above. information is repeated in this Section for clarity and to put prospective.

Since the early 1950's, there have been handbooks, pamphlets, and other publications, describing individual phases of fire service. There has always been a need for some type of publication, to encompass ladder designs, construction, maintenance, and testing. I have tried to include each subject in laymen's verbiage, to help reduce maintenance costs, and to maximize equipment longevity and performance.

"Third Party" Inspection Company inspectors can graduate from an NDT technical school with knowledge and accreditation, to perform as Level-2 Inspectors, in the fields for which they were trained and schooled. They will have knowledge of NDT equipment, and its usage, but still need field experience. NDT has three levels of training for expertise. The beginners are Level-1. This is where they read the procedure manuals and are still in training. In the field, these individuals can recognize a problem, but are not qualified to determine, make decisions, or statements, pro or con. After a period of field training and additional schooling, they will graduate to Level-2 status. At this level, the individuals should be able to do the inspection, make determinations, and file reports. The last is Level-3, where they are the instructors and trainers; and as a rule, they remain back at the home office, or at technical school. Level-3 may be Professional Engineers, graduates of a special class, and are usually accredited highly in their field of expertise. The Level-3 will answer questions that Level-2 may have and assist whenever they are needed. They will take reports the Level-2 Inspector has processed and review the procedures to be sure the Level-2 made the right determination, after the inspections. This should be the way it is in every situation. Financially stable "Third Party" Inspection Companies have the most opportunity to set the rules. This doesn't mean they are the most professional in the business, but money does talk. As I stated previously, the economy now dictates how a business can function financially, and who they can afford on payroll. First the "Third Party" Inspection Company starts with all the bells and whistles, gets accreditation and certification for wherever needed, and then is forced to cut personnel to the minimum. This is one reason why quality of inspections has changed to quantity of inspections.

We now have an idea what is required for the NDT part of apparatus inspections. NDT was added to fire apparatus inspections at first, to increase the average inspection report for the apparatus procedure. This addition would also eliminate most of the present inspectors that were not experienced in this field. There are drawbacks with NDT field inspecting that usually are not discussed with fire departments. There is magnetic particle inspection for steel, ultrasonic for pins and bolts, and dye penetrate for aluminum. These are the usual non-destructive tests performed in the field. Fire departments apparatus are usually cleaned and washed on a frequent schedule, to keep the apparatus looking good, and to remove most road dirt. Aerials and Platforms have lubrication applied and are cleaned on a cycle; or if there is a special condition, they must be cleaned more frequent. In the field, it is more difficult to magnetic particle inspect and get a true reading due to all the grease and paint on the ladder and on the boom sections. Using a handheld magnetic particle unit, iron powder, which come

Picture courtesy of C. Dewey

in a variety of colors, is sprinkled over weld areas, and the electrical magnet yolk ends can leave a burn indication on the equipment, which is difficult to remove from the ladder or boom sections. If the iron powder is not removed from the area previously checked, the iron powder will turn to red rust and discolor the ladder or boom sections. The advantage of using a handheld magnetic particle device is to check weld areas and structural ladder and boom components while in the field. If the inspector is doing his job correctly, he should look at each weld; and if there is a true visible indication of structural damage, or a possible weld crack, he will then clean around the area to confirm the results. Grease and paint can give false indications of structural damage. That's why one must be very careful when indicating a failure, as the apparatus would have to be removed from service and sent for repairs. Many units have been removed from service and sent for repair due to these false indications. This is costly to the fire departments, along with not having the apparatus in service for emergencies. The next step for checking the apparatus is the use of ultrasonic to check hinge pins and bolts for interior flaws, such as, cracks or broken pieces. This part of the inspection again forces the inspector, to visually inspect and ultrasonic all hinge pins and mounting bolts, to which there is access. Also, there is a drawback to this type of inspection. If the pins or bolts have external wear, the sonic beam will not pick it up on the screen. Holes have been drilled through pins and bolts which would show as a crack indication on the sonic screen, and must not be determined as a crack, until verified. Usually, pins are designed to withstand whatever force is given; but when cracked or broken, they should prohibit the use of that function before ultrasonic inspection is ever used. Mounting bolts for mainframe structures, turrets, turntable bearings, and any other structural components, may get loose and broken. Many of these bolts cannot be reached or checked due to inaccessibility. Most of the bolt heads will have strength indication marks on them, and that could restrict the ultrasonic transducer from obtaining a true reading on the screen. There is an advantage to this process, as it requires the inspector to look or touch and inspect each bolt that is accessible. Dye penetrate is the next type of NDT field structural defect inspection. This test verifies cracks in aluminum. Most of the time if there is a crack indication in aluminum welds or structures, the crack usually opens and is easy to view. Dye penetrate can stain the aluminum and should be used for special applications only.

Decades ago, there was a manufacturer that introduced into the fire service market, a heavy-duty welded aluminum ladder, and ladder platform. When "Third Party" Inspectors first began inspecting these new style ladders, they found many indications of what is called "crater cracks," at most of the weld endings or puddles. Aerial ladders with these indications were removed from service quicker than they were delivered. The manufacturer had to correct what had been written up by the "Third Party" Inspection Companies as structural cracks, per the NFPA pamphlet. After most of the aerials had been written up for repairs, the manufacturer sent out a bulletin to all the fire departments with these types of aerials, explaining that what was being written up was not an actual structural flaw, and how to handle each situation. At the beginning of manufacturing these welded aluminum aerials, there was a process for welding the structures together. What happened in this welding process was at the end of the weld, there would be a small "crater crack" in the weld material puddle. When this weld puddle cooled, the top would show a small surface "crater crack." These cracks were not structural flaws and could be removed with a simple pass of a die grinder; and the crack would disappear, and never return. This problem caused the manufacturer and fire departments much expense. The NFPA manual did not expound on this type of indication, so "Third Party" Inspectors had to consider these indications as structural defects, so they had the apparatus removed from service until these indications were repaired. After a short time, the manufacturer changed the welding procedure, and the problem was resolved.

The first aluminum aerial ladders manufactured had riveted members. There was no welding on the ladder sections. The first manufacturer of aluminum riveted ladder sections closed their doors in the early 1990's. There remains a manufacturer of aluminum riveted sections, who manufacturers a ladder and an aluminum riveted aerial platform. The rivet heads are large enough to ultrasonic for internal defects. With this type of structure, cracked welds were not a consideration. Cross members had to be checked for damage only, and rivets for loose mounting. The aluminum members could be verified for heat damage and proper hardness with a handheld test unit. There is one more NDT test that was introduced by a couple of "Third Party" Inspection Companies, and that test is called 'acoustic emission test.' This test uses sound waves and is supposed to find loose or weak areas in the boom or ladder sections. This type of inspection test uses a series of transducers, placed at various points on the ladder or boom sections, with results read at the receiver screen, while load testing the sections. Aluminum riveted aerial ladder sections make a lot of noise when extending and retracting. This test will show the areas from where the most noise occurs. Noise doesn't necessarily determine if there is weakness to the structural component, but only that an area should be checked. If aluminum shows heat damage or discoloring, the use of an aluminum hardness tester can verify if the material has weakened. This hardness tester also should be used to check any area where the transducers have detected noise. The hardness tester is the same type of equipment used in aluminum ground ladder inspection. Acoustic emission was primarily used to find the point of fatigue in fiberglass booms, and to verify whether the capacity rating of the boom was correct. The first acoustic emission tests were performed at utility power companies, on their fiberglass bucket truck booms, to verify capacity and boom fatigue. The testing agency would overstress the fiberglass booms, just to find out the maximum capacity the booms could stand, and at what point the fiberglass fibers would start to break down. At that point, the acoustic emission transducer would record the sound of the breaking fiber and pinpoint the area of failure of the fiberglass boom. This was not a smart way of thinking and testing. These booms were overstressed, left in service, many failed in the field, and injuries did occur. Caution by the utility companies was given to "Third Party" Inspection Companies to never over stress the material while performing this type test. The above NDT inspection procedures were required to have special levels of training, but the most efficient results were not in the field with all the elements against them, but at the manufacturing plant where the materials were in new and clean condition. To reiterate the benefits of this type of NDT inspection is that it should take more time to visually look over each item for damage and to slow down the inspector, thus creating a more thorough inspection.

The next part of NDT inspections does not require the levels of schooling and training to perform the actions. Load and structural apparatus stability tests can be performed with the use of load gauges, called "dynamometers." The other equipment used in the inspections, is dial indicators with magnetic bases. Dynamometers are scales with dials, indicating amount of load applied, or weight of the load being lifted. Dial indicators are what they imply. They have an indicator dial and a spring-loaded actuator on one end. They are primarily used in this inspection program, to check cylinder drifting. When load is applied, the needle on the gauge indicates the amount of cylinder compression, or cylinder drift, in very small increments. The dynamometer and dial indicator, work very well together, to reduce the possibility of overloading and false readings. There are two types of load tests; one is static or stationary loading; and the other test is free hanging and rotating the weight for stability. Static or stationary loading is primarily used for aerial ladders, positioned at the test elevation, and all sections to be fully extended.

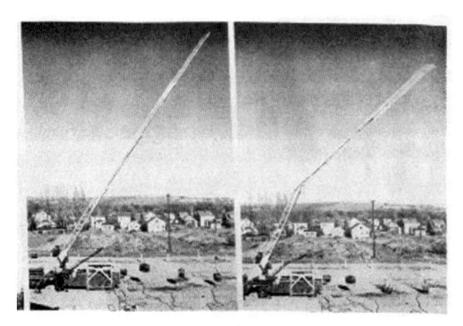

Picture courtesy of Fire Engineering

The ladder in the picture was one of the first tested metal ladders, following the NFPA recommendations. The ladder was tested by the fire department, with a hanging water weight. There was no way to prevent this failure with the weight in the air.

PTS placed aerial ladders in the same position as recommended by the NFPA. The only "difference" was when we attached the cable, we used a repelling rope with less than one percent stretch with strength of ten times the load test, yet the rope weighed only a few pounds. Then, we connected the come-a-long to the rope; and to achieve the correct load ratings, we attached the come-a-long to a dynamometer, attached to a positive, motionless, weight on the ground. With the come-a-long, we could add weight to the tip of the ladder and watch the dynamometer needle, for proper weight being applied, and for any creeping of the aerial. This is a safer way to add load weight, without shocking or swaying, an unsupported water load. Instead of marking the elevation cylinder shaft, we would install a dial indicator with a magnetic base to the cylinder shaft and touch the indicator needle to the cylinder housing or seal area, to measure drifting. Usually, if there is cylinder internal drifting, or oil compression, the dial indicator will show the amount of shaft retraction immediately. If drifting does not start until a load is applied to the aerial, the dynamometer needle will indicate if the ladder can or cannot support the load applied; and after checking the dial indicator for cylinder drifting, if there is excessive lowering, we immediately relieve the load, by releasing the come-a-long, with the weight attached, and determine if there is a problem. These aerials are not rotated and have a lesser load test than platforms or bucket trucks.

The next test is stability of an aerial platform or man lift. This test has a different load rating than on the aerial ladder. This test is performed as a stability test, along with cylinder drifting; and the load is rotated three hundred sixty degrees around the apparatus, in the most extreme position that the platform or man lift can obtain. On aerial platforms, and man lifts, used in utility departments, fire departments or other fields, the apparatus must be able to support at least one and one-half times the normal capacity of the platform or bucket, without losing stability. Basically, this test is to check supporting members, like outriggers on large units, and torsion bars on smaller units. If the one and one-half to one test load was placed inside the platform or basket, and the booms are positioned in the outermost reach position and then rotated to an unstable position; if the outriggers were not engineered to support the load test, and the platform did become unstable, the platform would lower to the ground, and could cause severe damage. If the load test weight was attached and supported under the center of the platform or basket, but very close to the ground; when the boom controls were operated and the booms rotated to the most unstable position, if the platform started to lower toward the ground, the weight would immediately release; and the platform could be saved from overloaded failure. Many platforms were found unable to support the overload test at the maximum position.

Another great use for the dynamometer in the field is to reset overload protector alarms that were installed on some aerial apparatus, and to verify if they are correctly set to the recommendations of the manufacturer. Using an anchor, or weight that exceeds the overload alarm rating, position it on the ground under the horizontal fully extended ladder tip. Next, attach the dynamometer to the stationary weight. Attach a rope or cable to the tip of the extended ladder. Connect the rope or cable to the come-a-long and attach the other end of the come-a-long to the dynamometer. When everything is ready for a straight vertical pull, using the come-a-long, gently apply pressure to the ladder tip, while watching the dial needle on the dynamometer. Apply the recommended load for that unit; and the alarm, if properly adjusted, should sound. Do not exceed the rated load. If the alarm does not sound, adjust the alarm instrument to sound at the desired weight, but not to exceed the rated capacity. These special load alarms do get out of adjustment, and this could present an overloading condition to the ladder sections, and a false sense of security for the operator. Have the manufacturer or representative check the alarm on a normal schedule.

I have explained what part the NDT portion of a "Third Party" Apparatus Inspection curtails. It still is not the ultimate answer to a complete thorough safety inspection and does have loopholes in the field inspections.

We have just gone through weld indications, pins for damage, bolts for looseness and possible breaks, cylinders for leaks, outriggers for stability, and noise for structural weak areas in boom and ladder sections. Now, if all these areas of inspection and testing prove satisfactory, the unit should be ready to certify for fire service. "Right?" Only if the mechanic or in-house fire department inspector has performed all his or her duties in maintaining the apparatus, then the apparatus is ready for service, without the possibility of failure of a component while on an emergency call. While doing regular maintenance on the apparatus, you should keep in mind the following items:

When I begin an apparatus inspection, my first thought is: If the vehicle cannot make it to the emergency, the unit mounted to the chassis would be dysfunctional. So, here is where I initiate every apparatus inspection. The first thing to notice is chassis for damage, and whether all the tires look normal. Then take the tire pressure, because these large tires won't always show they are flat, especially on the duals. While checking the air pressure, notice the wheel lug nuts. Some may be missing, or they all appear to be there; but are they properly tightened? Torque each lug nut with your hands, and you will find many have loosened. Once around the chassis after tire inspection is the beginning of the underside of chassis inspection. Starting with the very front of the chassis there will be mounting bolts for many components that may loosen, break off, or are just missing. The front bumper and siren mounting, and body bolts should be checked; the steering box assembly checked for mounting and grease leaks; wiring and any hoses are checked for wear and chaffing. Moving back underneath the chassis, the front brakes, and drums, should be checked for wear and structural cracks. There is an inspection shield at the brake drums that should be removed to check pads for wear and brake drums for stress cracks. Check the front springs for broken leaf's and broke or missing spring clips which hold the spring leaves in place. Check shock absorbers for damage and mounting. Check the brake slack adjusters, while working your way back to the rear of the chassis; and after all adjusters are properly mounted, check the stroke for proper setting of the brakes. The rear brakes should set quicker than the front, to help eliminate wear and surging on the front brakes. Visually check the chassis frame while moving toward the rear for structural damage, missing bolts, loose mounting of components, and wiring and hoses rubbing on frame members. Many chassis frame flanges have some type of rubber hose, or some other component, to cushion components from chaffing on the flange edges. These cushions will loosen, and sometimes fall off the flange, causing hoses or wiring covers and braids to wear or chaff on the frame flanges. Driveline universal joints should be checked for bearing wear and movement, also for missing mounting bolts, and proper lubrication. This doesn't take long to view, and checking for wear of components, improves the inspection. The apparatus mainframe, mounted to the chassis frame, can shift and elongate frame mounting holes; sometimes mounting bolts will break and disappear, and many will loosen. If the torque on the mainframe bolts is not to be checked at this service, simply check the bolts by using your hand to check for looseness. Perform a quick visual on the welds on the mainframe and other components for indications of wear or cracking, while checking the frame mounting bolts. Now, we have reached the center of the chassis, where the hydraulic components are usually found. The hydraulic reservoir, filter, pto-pump, and hoses that send the hydraulic power to the upper controls, are in this general area. If the hydraulic oil is not sent out for a chemical analysis test, make sure the oil is at proper level and not contaminated with moisture. The hoses and filter should be checked for leaks, and hoses for chaffing. Hydraulic oil and hydraulic filters last much longer than the normal oil and filters for the apparatus engine. Continuing toward the rear of the chassis, the exhaust might have leakage or damage to the muffler. This is easy to spot; and if damages are noticed, they should be repaired or replaced. There will be many chassis body supports mounted to the chassis frame. Bolts will loosen easily, and body compartments can shift. Further back, the rear springs should be viewed for broken leaves, and broken or missing spring clips. Check the rear brake canisters for mounting and stroke to properly set the brakes. Rear wheel inspection plates should be removed, brakes checked for wear, and drums for cracks and broken parts. The ladder or boom travel support can be mounted behind the cab, or at the rear of the chassis, and mounted to frame members. Check the support for stress cracks and bolts for looseness. We have now done a quick undercarriage inspection and are at the back of the chassis. The only items left, before the apparatus is road ready, is to check all the lights, turn signals, back up alarm, siren, and emergency items. By checking most of the above items on a regular schedule, or while waiting for the oil to drain and changing filters, visual inspections can lower maintenance costs that would multiply, if any of the above were left unnoticed.

Cleaning and lubricating aerial ladders, is different than cleaning aerial platforms. Ladder sections have guides that require periodic lubrication and cleaning, more than solid boom sections on aerial platforms. In the early years of aerial ladder manufacturing, there was only one type of lubrication recommended for ladder guides. This was called lubriplate grease and was the best lubricant at that time. The problem with lubriplate grease is it would harden after exposed to dirt and grime. With the dirt and the hardening, this lubricant would cease up the rollers, causing them to wear off the apex, and stop the function of the rollers and aerial functions. Cleaning the ladder sections became a chore, and much of the grease usually had to be pressure steamed off or scrapped off with a blade. This lubrication remained a recommendation by the NFPA, and there never seemed to be a priority for a better lubricant. In the 1980's, multi-colored silicone grease was new on the market, was tried, and the results were outstanding. This type of lubricant would put a lighter film on the guides and would never seem to harden from dirt and the elements. Excessive grease on the ladder sections was a very common occurrence. There was excessive grease on ladder side rails and in areas that should never need lubrication. On aerial ladder guides, if lubrication is lightly applied to the top edge of the inside lower guide of the base ladder, then the top of the lower outside guides and the top of the inside lower guide edges of the mid sections, that will be all that is required for the base ladder and mid extension sections guides. For the fly section, only a light lubrication needs to be applied to the top of the lower outside guide. On all the extension sections (only), a light lubrication should be applied to the underside of the rails, so they will slide smoothly over the rollers, or guide blocks. This should be all the lubrication the ladder sections need. Cleaning the ladder sections without steam cleaning, and damaging the paint, or using blades, is a very simple process. For a one-hundred-foot aerial ladder, take two workers, two six footstep ladders, a bag of cleaning rags, and six cans of brake cleaner (used for brake drums cleaning), or equivalent. Extend the ladder sections fully horizontal. Support the tip end of the fly section, especially for light-duty aerials. Spray the cleaner on the guides, and wipe off the old, contaminated grease. Brake cleaner has trichloroethylene, a cleaning agent that quickly dries and leaves no residue.

It may be a little toxic, so care must be taken. Cleaning the ladder sections this way should only take about two hours.

This general chassis visual inspection is also for all pieces of emergency equipment, as in pumpers, tankers, and command units. Pumpers have their own performance tests for checking pump accuracy, and need a procedure for checking undercarriage, as shown above, to make sure all hoses and fittings, along with emergency tools and equipment required, are in place and maintained. The other equipment has similar description, and they are inspected in the same manner adaptable to each piece of equipment.

There is one item on all emergency equipment, and in the firehouse, that gets over-looked occasionally; and that item is the fire extinguisher. Larger fire departments usually have a fire extinguisher company check all the extinguishers on a regular basis; but it is surprising how many are overlooked and have expired certification dates.

APPRECIATION:

Fire Service ll, "The Other Side's" focus encompasses safety inspections, maintenance, education, a history of Professional Testing Systems (PTS) as a "Third Party" Inspection Company and illustrates how PTS played a role in design and inspecting firefighting apparatus and equipment for safety. If I can provide any apparatus operator, fireman or mechanic with one idea on how he or she can improve on their fire service procedures or their safety inspections, from my experiences, would be the pinnacle of my life. This is my primary objective for writing these articles. In twenty years of PTS inspections, the number of overlooked items, or damage caused to fire equipment by any PTS company inspector, could have been counted on one hand; and thank God, all incidents were very minor. To err is human, but much care and patience is required. We were not always perfect in our efforts, but we always gave over 100%.

Looking back at some of the failures and problems that resulted with various types of apparatus from early manufacture to present day. In the early years of manufacturing, wood was the primary product used to develop ground ladders and chassis mounted aerial ladders. Trial and error resulted in many failures but led the way to bigger and better products. Later wood was changed to metal, and a much stronger ladder emerged in the fire industry. Early 1980's, there were forty-eight cities in the U.S. and Canada that had fifty-three aerial failures, some that resulted in injury and fatalities. This does not include the many unreported incidents and unknown accidents throughout those years. Many of the more serious incidents were investigated by manufacturers, engineers, graduates from Fire Science Schools, or other individuals who had some type of fire apparatus training. Results were a mixed bag as to operator neglect or apparatus failure.

Fire apparatus is a special field with many challenges and rewarding accomplishments. Our goal was to make sure the equipment was following manufacturer's standards and in the best condition for fire service at the time of the inspection. Driving across the U.S. two or three times a year, puts a lot of miles on the vehicle, and on the driver. The office would make out a schedule in a designated area of the U. S., to group fire departments, and to make schedules for apparatus inspections. There were very few departments that would not greet our visit in a pleasant way. Whenever lunch was served, we were always a guest; and if the inspections lasted into the night, dinner was always offered, in most cases. Most fire department cooks should be considered chefs, as the food they served was always outstanding. I remember one occasion when we were served venison for dinner. I exclaimed to the cook how delicious it was, and then had to ask the question, "Who shot the deer?" They didn't want to tell me that a truck had just hit the deer hours earlier, and it was roadkill. That happened all the time. Working in larger cities at their shops, we would have to stop inspection when the shop doors closed, and work with their schedules. When we worked in the New Orleans area, the firefighters would go out of the way to have a large crawfish boil. The first boil we went to, was a learning event. Everyone at the table ended up with large piles of crawfish shells stacked in front of them, and we only had a few, because we peeled each one individually. That does not happen anymore. The Northeast always had a great feast, mostly of some type of fish or shellfish; it was always worth going to that part of the country. When a fire department respects what you are doing for them, they seem to go out of their way to help and assist.

PTS has been very blessed, for the past twenty years of service, to have no serious failures of aerial devices in this profession. These are very high dollar pieces of equipment and must be treated accordingly.

In December 1998, it was 5:00 o'clock in the evening; I was sitting in stopped traffic in front of the San Diego Airport on highway 5, in a new rented vehicle, waiting to return to my hotel room, after working at the first fire department where I had worked for more than twenty years. I glanced in my rear-view mirror and noticed a truck driving in my stopped lane, heading toward me and going very fast.

After my automobile accident in San Diego, it was time to retire PTS and enjoy all the memories and experiences, that the volunteer and paid fire department personnel, had shared with me and all the inspectors employed by PTS throughout the years. PTS was a small company that took seriously the responsibilities of aerial inspections. It was difficult to expand with more crews, because of depth than must be followed to give a 100% inspection. There is no school, or university that teaches, or even has the books with the knowledge, that we have learned from this life experience, while working in the field every day. To reiterate, firefighters and the men and women in the fire service, to include emergency medical personnel, are a very special breed, dedicated to give the world, what they do best. The personnel, and manufacturers in "The Other Side," no matter what the economy trend is, made sure the equipment, training, maintenance, and safety inspections are always of the best quality for the firefighting industry. For all the manufacturers of fire equipment, the twentieth century was full of changes. They were upgrading, manufacturing, and developing safer and more durable equipment and apparatus for fire department use. The twenty-first century, with all the new innovations and automation, will set the pace for future development of new upgraded technical firefighting and rescue equipment and apparatus.

PTS closed its doors in 1999.

Thanks to all the fire personnel that assisted me and my inspectors through all the years PTS was able to assist fire departments everywhere.